A

CATECHISM

OF THE

STEAM ENGINE,

ILLUSTRATIVE OF THE

SCIENTIFIC PRINCIPLES UPON WHICH ITS OPERATION DEPENDS; AND THE PRACTICAL DETAILS OF ITS STRUCTURE, IN ITS APPLICATION TO MINES, MILLS, STEAM NAVIGATION AND RAILWAYS.

With various Suggestions of Improvement.

BY JOHN BOURNE, C. E.

A NEW EDITION.

NEW-YORK:
D. APPLETON & COMPANY,
346 & 348 BROADWAY.
M.DCCC.LIV.

PREFACE.

The present work is not intended as a substitute for the quarto Treatise on the Steam Engine, which I lately published, but is rather to be regarded as an introduction and in some measure also as a supplement to that work. Notwithstanding the existence, therefore, of the larger Treatise, it appeared to me that a work upon the steam engine, which in a moderate compass should give an outline of the whole subject in its practical aspect, would still be of much utility. There are no doubt many compendiums already existing which profess to accomplish this object; but I have not met with any which were calculated to satisfy even the most moderate expectations. Most of them are mere compilations from theoretical authors, and abound even with scientific errors, whilst indicating the absence of any practical acquaintance with the subject; so that they possess but slender claims upon the attention of the engineer, or indeed of any one desirous of obtaining accurate information on the subject of which they treat. I hold it to be the first quality of an introductory work that it should at least be sound—that the doctrines it inculcates, and the lessons it conveys, shall not all have to be unlearned again at a subsequent stage of progress — and whatever be its other characteristics, I believe that the present work will at least be found to conform to this standard of utility. It embodies, I believe, the best information now existing upon the subjects of which it treats—not taken from books, nor deduced from mere theoretical considerations, but derived from my own practice or from the personal communications of the most experienced engineers of the present time.

JOHN BOURNE.

A CATECHISM

OF

THE STEAM ENGINE.

PART I.

MECHANICAL PRINCIPLES.

1. *Q.*—What is meant by a vacuum?

A.—A vacuum means an empty space; a space in which there is neither water nor air, nor any thing else that we know of.

2. *Q.*—Wherein does a high pressure differ from a low pressure engine?

A.—In a high pressure engine the steam, after having pushed the piston to the end of the stroke, escapes into the atmosphere, and the impelling force is therefore that due to the difference between the pressure of the steam and the pressure of the atmosphere. In the condensing engine the steam, after having pressed the piston to the end of the stroke, passes into the condenser, in which a vacuum is maintained, and the impelling force is that due to the difference between the pressure of the steam above the piston, and the pressure of the vacuum beneath it, which is nothing; or, in other words, you have then the whole pressure of the steam urging the piston, consisting of the pressure shown by the safety valve on the boiler, and the pressure of the atmosphere besides.

3. *Q.*—How can the pressure of a vacuum be said to be nothing, when the existence of a vacuum occasions a pressure of 15 lbs. on the square inch?

A.—Because it is not the vacuum which exerts this pressure, but the atmosphere, which, like a head of water, presses on every thing immerged beneath it. A head of water, however, would not press down a piston, if the water were admitted on both its sides; for an equilibrium would then be established, just as a balance retains its equilibrium though an equal weight be added to each scale; but take the weight out of one scale, or empty the water from one side of the piston, and motion or pressure is produced; and in like manner pressure is produced on a piston by admitting steam or air upon the one side, and withdrawing the steam or air from the other side. It is not, therefore, to a vacuum, but rather to the existence of an unbalanced plenum, that the pressure made manifest by exhaustion is due, and every one knows that a vacuum of itself would not work an engine.

4. *Q.*—How is the vacuum maintained in a condensing engine?

A.—The steam, after having performed its office in the cylinder, is permitted to pass into a vessel called the condenser, where a shower of cold water is discharged upon it. The steam is condensed by the cold water, and falls in the form of hot water to the bottom of the condenser. The water, which would else be accumulated in the condenser, is continually being pumped out by a pump worked by the engine.

5. *Q.*—If a vacuum be an empty space, and there be water in the condenser, how can there be a vacuum there?

A.—There is a vacuum above the water, the water being only like so much iron or lead lying at the bottom.

6. *Q.*—Is the vacuum in the condenser a perfect vacuum?

A.—Not quite perfect; for the cold water entering for the purpose of condensation is heated by the steam, and emits a vapor of a tension represented by about three inches of mercury; that is, when the common barometer stands at 30 inches, a barometer with the space above the mercury communicating with the condenser will stand at about ·27 inches.

7. *Q.*—Is a barometer sometimes applied to the condensers of steam engines?

A.—Yes; and it is called the vacuum gauge, because it shows the degree of perfection the vacuum has attained. Another gauge, called the steam gauge, is applied to the boiler, which indicates the pressure of the steam by the height to which the steam forces mercury up a tube.

8. *Q.*—Can a condensing engine be worked with a pressure less than that of the atmosphere?

A.—Yes, if once it be started; but it will be a difficult thing to start an engine, if the pressure of the steam be not greater than that of the atmosphere. Before an engine can be started, it has to be blown through with steam to displace the air within it, and this cannot be effectually done if the pressure of the steam be very low. After the engine is started, however, the pressure in the boiler may be lowered, if the engine be lightly loaded, until there is a partial vacuum in the boiler. Such a practice, however, is not to be commended, as the gauge cocks, which are cocks applied for telling the height of water within the boiler, become useless when there is a partial vacuum in the boiler; inasmuch as, when they are opened, the water will not rush out, but air will rush in. It is impossible, also, under such circumstances, to blow out any of the sediment collected within the boiler, which, in the case of the boilers of steam vessels, requires to be done every two hours. In some cases, in which the boiler applied to an engine is of inadequate size, the pressure within the boiler will fall spontaneously to a point considerably beneath that of the atmosphere; but it is preferable in such cases partially to close the throttle valve in the steam pipe, whereby the issue of the steam is resisted; and the pressure in the boiler is thus maintained, though the cylinder only receives its former supply.

9. *Q.*—In what way would you class the various kinds of condensing engines?

A.—Into single acting, rotative, and rotatory engines. Single acting engines are engines without a crank, such as are used for pumping water. Rotative engines are engines provided with a crank, by means of which a rotative motion is produced; and in this important class stand marine and mill engines, and all engines, indeed, in which the rectilinear motion of the piston is changed into a circular motion. In rotatory engines the steam acts at once in the production of circular motion, either upon a revolving piston or otherwise, but without the use of any intermediate mechanism, such as the crank, for deriving a circular from a rectilinear motion. Rotatory engines have not hitherto been successful, so that only the single acting or pumping engine, and the double acting or rotative engine, can be said to be in actual use. For some purposes, such, for example, as forcing air into furnaces for smelting iron, double acting engines are employed, which are nevertheless unfurnished with a crank; but engines of this kind are not sufficiently numerous to justify their classification as a distinct species, and, in general, those engines may be considered to be single acting, by which no rotatory motion is imparted.

10. *Q.*—Is not the circular motion derived from a cylinder engine very irregular, in consequence of the unequal leverage of the crank at the different parts of its revolution?

A.—No; rotative engines are generally provided with a fly-wheel to correct such irregularities by its momentum; but where two engines, with their respective cranks set at right angles, are employed, the irregularity of one engine corrects that of the other with sufficient exactitude for many purposes. In the case of marine and locomotive engines, a fly-wheel is not employed; but for cotton spinning, and other purposes requiring great regularity of motion, its use with common engines is indispensable, though it is not impossible to supersede the necessity by new contrivances.

11. *Q.*—You implied that there is some other dif-

ference between single acting and double acting engines, than that which lies in the use or exclusion of the crank?

A.—Yes; single acting engines act in only one way by the force of the steam, and are returned by a counter weight; whereas double acting engines are urged by the steam in both directions. Engines, as I have already said, are sometimes made double acting though unprovided with a crank; and there would be no difficulty in so arranging the valves of ordinary pumping engines, as to admit of this action; for the pumps might be contrived to raise water both by the upward and downward stroke, as indeed in some mines is already done. But engines without a crank are almost always made single acting, perhaps from the effect of custom as much as from any other reason, and are usually spoken of as such, though it is necessary to know that there are some deviations from the usual practice.

12. *Q.*—With what velocity does air rush into a vacuum?

A.—With the velocity which a body would acquire by falling from the height of a homogeneous atmosphere. The weight of air being known, as well as the pressure it exerts on the earth's surface, it becomes easy to tell what height a column of air an inch square, and of the atmospheric density, would require to be to weigh 15 lbs. The height would be 27,818 feet.

13. *Q.*—And what velocity would the fall of a body from such a height produce?

A.—About 1,338 feet per second. All bodies fall

with the same velocity when there is no resistance from the atmosphere, as is shown by the experiment of letting fall a feather and a guinea from the top of a tall exhausted receiver, when they reach the bottom at the same time. The velocity of falling bodies is one that is accelerated uniformly, according to a known law; and when the height from which a body falls is given, the velocity acquired at the end of the descent can be easily computed. The square root of the height in feet multiplied by 8·021 will give the velocity.

14. *Q.*—But the velocity in what terms?

A.—In feet per second. The distance through which a body falls by gravity in one second is $16\frac{1}{12}$ feet; in two seconds, $64\frac{4}{12}$ feet; in three seconds, $144\frac{9}{12}$ feet; in four seconds, $257\frac{4}{12}$ feet, and so on. If the number of feet fallen through in one second be taken as unity, then the relation of the times to the spaces will be as follows:—

Number of seconds	1	2	3	4	5	6	&c.
Units of space passed through	1	4	9	16	25	36	

so that it appears that the spaces passed through by a falling body are as the squares of the times of falling. The velocity acquired by a falling body *at the end* of the 1st second is $32\frac{1}{6}$ feet per second, at the end of the 2d second $64\frac{2}{6}$ feet, at the end of the 3d, $96\frac{3}{6}$ feet, at the end of the 4th, $128\frac{4}{6}$ feet, and so on. These numbers proceed in the progression 1, 2, 3, 4, &c., so that it appears that the velocities acquired by a falling body at different points, are simply as the times of falling. But if the velocities be as the times, and the total space passed through be as the squares of the times, then the total space passed through must be as the squares of the velocity; and as the *vis viva* or mechanical power inherent in a falling body, of any given weight, is measurable by the height through which it descends, it follows that the *vis viva* is proportionate to the square of the velocity. Of two balls therefore, of equal weight, but one moving twice as fast as the other, the faster ball has four times the energy or mechanical force accumulated in it that the slower ball has. If the speed of a fly-wheel be doubled, it has four times the energy it possessed before—such energy being measurable by a reference to the height through which a body must have fallen, to acquire the velocity given.

15. *Q.*—By what considerations is the momentum proper for the fly-wheel of an engine determined?

A.—By a reference to the power produced every half-stroke of the engine, joined to the consideration of what relation the energy of the fly-wheel rim must have thereto, to keep the irregularities of motion within the limits which are admissible. It is found in practice, that when the power resident in the fly-wheel rim, when the engine moves at its average speed, is from two and a half to four times greater than the power generated by the engine in one half-stroke — the variation depending on the momentum inherent in the machinery the engine has to drive and the equability of motion required — the engine will work with sufficient regularity far all ordinary purposes.

16. *Q.*—What do you understand by the centre of gravity of a body?

A.—That point within it, in which the whole of the weight may be supposed to be concentrated, and which continually endeavors to gain the lowest possible position. A body hung in the centre of gravity will remain at rest in any position.

17. *Q.*—What do you understand by centrifugal and centripetal forces?

A.—By centrifugal force, I understand the force with which a revolving body tends to fly from the centre; and by centripetal force, I understand any force which draws it to the centre, or counteracts the centrifugal tendency. In the conical pendulum, or steam engine governor, which consists of two metal balls suspended on rods hung from the end of a vertical revolving shaft, the centrifugal force is manifested by the divergence of the balls when the shaft is put into revolution; and the centripetal force, which in this instance is gravity, predominates so soon as the velocity is arrested; for the arms then collapse and hang by the side of the shaft.

18. *Q.*—What is meant by the centre of gyration?

A.—The centre of gyration is that point in a revolving body in which the whole momentum may be conceived to be concentrated, or in which the whole effect of the momentum resides. If the ball of a governor were to be moved in a straight line, the momentum might be said to be concentrated at the centre of gravity of the ball; but inasmuch as, by its revolution round an axis, the part of the ball furthest removed from the axis moves quicker than the part nearest to it, the momentum cannot be supposed to be concentrated at the centre of gravity, but at a point further removed from the central shaft, and that point is what is called the centre of gyration.

19. *Q.*—What is the centre of oscillation?

A.—The centre of oscillation is a point in a pendulum or any swinging body, such, that if all the matter of the body were to be collected into that point the velocity of its vibration would remain unaffected. It is in fact the mean distance from the centre of suspension of every atom, in a ratio which happens not to be an arithmetical one. The centre of oscillation is always in a line passing through the centre of suspension, and the centre of gravity.

20. *Q.*—By what circumstance is the velocity of vibration of a pendulous body determined?

A.—By the length of the suspending rod only, or, more correctly, by the distance between the centre of suspension and the centre of oscillation. The length of the arc described does not signify, as the times of vibration will be the same, whether the arc be the fourth or the four hundredth of a circle; or at least they will be nearly so, and would be so exactly, if the curve described were a portion of a cycloid. In the pendulums of clocks, therefore, a small arc is preferred, as there is, in that case, no sensible deviation from the cycloidal curve, but in other respects the size of the arc does not signify.

21. *Q.*—If then the length of a pendulum be given, can the number of vibrations in a given time be determined?

A.—Yes; the time of vibration bears the same relation to the time in which a body would fall through a space equal to half the length of the pendulum, that the circumference of a circle bears to its diameter. The number of vibrations made in a given time by pendulums of different lengths, is inversely as the square roots of their lengths.

22. *Q.*—Then when the length of the second's pendulum is known, the proper length of a pendulum to make any given number of vibrations in the minute can readily be computed?

A.—Yes; the length of the second's pendulum being known, the length of another pendulum, required to perform any given number of vibrations in the minute, may be obtained by the following rule: multiply the square root of the given length by 60, and divide the product by the given number of vibrations per minute; the square of the quotient is the length of pendulum required. Thus if the length of a pendulum were required that would make 70 vibrations per minute in the latitude of London, then $\frac{\sqrt{39 \cdot 1393} \times 60}{70} = 5 \cdot 363^2 = 28 \cdot 75$ in., which is the length required.

23. *Q.*—What measures are there of the centrifugal force of bodies revolving in a circle?

A.—The centrifugal force of bodies revolving in a circle increases as the diameter of the circle, if the number of revolutions remain the same. If there be two fly-wheels of the same weight, and making the same number of revolutions per minute, but the diameter of one be double that of the other, the larger will have double the amount of centrifugal force. The centrifugal force of the *same wheel,* however, increases as the square of the velocity; so that if the velocity of a fly-wheel be doubled, it will have four times the amount of centrifugal force.

24. *Q.*—Can you give a rule for determining the centrifugal force of a body of a given weight moving with a given velocity in a circle of a given diameter?

A.—Yes. If the velocity in feet per second be divided by 4·01, the square of the quotient will be four times the height in feet from which a body must have fallen to have acquired that velocity. Divide this quadruple height by the diameter of the circle, and the quotient is the centrifugal force in terms of the weight of the body; so that multiplying the quotient by the actual weight of the body, we have the centrifugal force in pounds or tons. Another rule is to multiply the square of the number of revolutions per minute by the diameter of the circle in feet, and to divide the product by 5,870. The quotient is the centrifugal force in terms of the weight of the body.

25. *Q.*—Can you explain how it comes that the length of a pendulum determines the number of vibrations it makes in a given time?

A.—Because the length of the pendulum determines the steepness of the circle in which the body moves, and it is obvious, that a body will descend more rapidly over a steep inclined plane, or a steep arc of a circle, than over one in which there is but a slight inclination. The impelling force is gravity, which urges the body with a force proportionate to the distance descended; and if the velocity due to the descent of a body through a given height be spread over a great horizontal distance, the speed of the body must be slow in proportion to the greatness of that distance. It is clear, therefore, that as the length of the pendulum determines the steepness of the arc, it must also determine the velocity of vibration.

26. *Q.*—If the motions of a pendulum be dependent on the speed with which a body falls, then a certain ratio must subsist between the distance through which a body falls in a second, and the length of the second's pendulum?

A.—And so there is; the length of the second's pendulum at the level of the sea in London, is 39·1393 inches, and it is from the length of the second's pendulum that the space through which a body falls in a second has been determined. As the time in which a pendulum vibrates is to the time in which a heavy body falls through half the length of the pendulum, as the circumference of a circle is to its diameter, and as the height through which a body falls is as the square of the time of falling, it is clear that the height through which a body will fall, during the vibration of a pendulum, is to half the length of the pendulum as the square of the circumference of a circle is to the square of its diameter; namely, as 9·8696 is to 1, or it is to the whole length of the pendulum as the half of this, namely, 4·9348 is to 1; and 4·9348 times 39·1393 in. is $16\frac{1}{12}$ ft. very nearly, which is the space through which a body falls by gravity in a second.

27. *Q.*—Are the motions of the conical pendulum or governor reducible to the same laws which apply to the common pendulum?

A.—Yes; the motion of the conical pendulum may

be supposed to be compounded of the motions of two common pendulums, vibrating at right angles to one another, and one revolution of a conical pendulum will be performed in the same time as two vibrations of a common pendulum, of which the length is equal to the vertical height of the point of suspension above the plane of revolution of the balls. A steam engine governor may, it is true, be driven round with any speed, and therefore it may be supposed that the length of the arms cannot affect the time of revolution; but as the speed is increased the balls expand, and the height of the cone described by the arms is diminished, until its vertical height is such that a pendulum of that length would perform two vibrations for every revolution of the governor. If, therefore, a certain expansion of the balls be desired, and a certain length be fixed upon for the arms, so that the vertical height of the cone is fixed, then the speed of the governor must be such, that it will make half the number of revolutions in a given time that a pendulum equal in length to the height of the cone would make of vibrations. The rule is, multiply the square root of the height of the cone in inches by 0·31986, and the product will be the right time of revolution in seconds. If the number of revolutions and the length of the arms be fixed, and it is wanted to know what is the diameter of the circle described by the ball, you must divide the constant number 187·68 by the number of revolutions per minute, and the square of the quotient will be the vertical height in inches of the centre of suspension above the plane of the ball's revolution. Deduct the square of the vertical height in inches from the square of the length of the arm in inches, and twice the square root of the remainder is the diameter of the circle in which the centres of the balls revolve.

28. *Q.*—Cannot the operation of a governor be deduced merely from the consideration of centrifugal and centripetal forces?

A.—It can, and by a very simple process. The horizontal distance of the arm from the spindle divided by the vertical height, will give the amount of centripetal force, and the velocity of revolution requisite to produce an equivalent centrifugal force may be found by multiplying the centripetal force of the ball in terms of its own weight by 70,440, and dividing the product by the diameter of the circle made by the centre of the ball in inches: the square root of the quotient is the number of revolutions per minute. By this rule you fix the length of the arms, and the diameter of the base of the cone, or, what is the same thing, the angle at which it is desired the arms shall revolve, and then you make the speed or number of revolutions such that the centrifugal force will keep the balls in the desired position.

29. *Q.*—Does not the weight of the balls affect the question?

A.—Not in the least; each ball may be supposed to be made up of a number of small balls or particles, and each particle of matter will act for itself. Heavy balls attached to a governor are only requisite to overcome the friction of the throttle valve which shuts off the steam, and of the connections leading thereto. Though the weight of a ball increases its centripetal force, it increases its centrifugal force in the same proportion,

30. *Q.*—What do you understand by the mechanical powers?

A.—The mechanical powers are certain contrivances, such as the wedge, the screw, the inclined plane, and other elementary machines, which convert a small force acting through a great space into a great force acting through a small space. In the school treatises on mechanics, a certain number of these devices are set forth as the mechanical powers, and each separate device is treated as if it involved a separate principle; but not a tithe of the contrivances which accomplish the stipulated end are represented in these learned works, and there is no very obvious necessity for considering the principle of each contrivance separately when the principles of all are one and the same. Every pressure acting with a certain velocity, or through a certain space, is convertible into a greater pressure acting with a less velocity, or through a smaller space; but the quantity of mechanical force remains unchanged by this transformation, and all that implements called mechanical powers accomplish, is to effect this transformation.

31. *Q.*—Is there no power gained by the lever?

A.—Not any; the power is merely put into another shape, just as the contents of a hogshead of porter are the same, whether they be let off by an inch tap or by a hole a foot in diameter. There is a greater gush in the one case than in the other, but it will last a shorter time: when a lever is used there is a greater force exerted, but it acts through a shorter distance. It requires just the same expenditure of mechanical power to lift 1 lb. through 100 ft., as to lift 100 lbs.

through one foot. A cylinder of a given cubical capacity will exert the same power by each stroke, whether the cylinder be made tall and narrow, or short and wide; but in the one case it will raise a small weight through a great height, and in the other case, a great weight through a small height.

32. *Q.*—Is there no loss of power by the use of the crank?

A.—Not any. Many persons have supposed that there was a loss of power by the use of the crank, because at the top and bottom centres it is capable of exerting little or no power; but at those times there is little or no steam consumed, so that no waste of power is occasioned by the peculiarity. Those who imagine that there is a loss of power caused by the crank, confuse themselves by confounding the vertical with the circumferential velocity. If the circle of the crank, be divided by any number of equidistant horizontal lines, it will be obvious that there must be the same steam consumed, and the same power expended, when the crank-pin passes from the level of one line to the level of the other, in whatever part of the circle it may be, those lines being indicative of equal ascents or descents of the piston. But it will be seen that the circumferential velocity is greater with the same expenditure of steam when the crank-pin approaches the top and bottom centres; and this increased velocity exactly compensates for the diminished leverage, so that there is the same power given out by the crank in each of the divisions.

33. *Q.*—Have no plans been projected for gaining power by means of a lever?

A.—Yes, many plans—some of them displaying much ingenuity, but all displaying a complete ignorance of the first principles of mechanics, which teach that power cannot be gained by any multiplication of levers and wheels. I have occasionally heard persons say: "You gain a great deal of power by the use of a capstan: why not apply the same resource in the case of a steam vessel, and increase the power of your engine by placing a capstan motion between the engine and paddle-wheels?" Others I have heard say: "By the hydraulic press you can obtain unlimited power: why not then interpose a hydraulic press between the engines and the paddles?" To these questions the reply is sufficiently obvious. Whatever you gain in force you lose in velocity; and it would benefit you little to make the paddles revolve with ten times the force, if you at the same time caused them to make only a tenth of the number of revolutions. You cannot, by any combination of mechanism, get increased force and increased speed at the same time, or increased force without diminished speed; and it is from the ignorance of this inexorable condition, that such myriads of schemes for the realization of perpetual motion, by combinations of levers, weights, wheels, quicksilver, cranks, and other mere pieces of inert matter, have been propounded. Any such combination can never increase power, nor diminish it either, except by friction. Power is not measurable by force, but by force and velocity combined.

34. *Q.*—What is friction?

A.—Friction is the resistance experienced when

one body is rubbed upon another body, and is the result of the natural attraction bodies have for one another, and of the interlocking of the impalpable asperities upon the surfaces of all bodies, however smooth. There is also, no doubt, some electrical action involved in its production, not yet recognized nor understood. When motion in opposite directions is given to smooth surfaces, the minute asperities of one surface must mount upon those of the other, and both will be abraded and worn away: in which act power must be expended. The friction of smooth rubbing substances is less when the composition of those substances is different, than when it is the same, the particles being supposed to interlock less when the opposite prominences or asperities are not coincident.

35. *Q.*—Does friction increase with the extent of rubbing surface ?

A.—No ; the friction, so long as there is no violent heating or abrasion, is simply in the proportion of the pressure keeping the surfaces together, or nearly so. It is, therefore, an obvious advantage to have the bearing surfaces of steam engines as large as possible, as there is no increase of friction by extending the surface, while there is a greater increase in the durability. When the bearings of an engine are made too small, they very soon wear out.

36.—*Q.* Does friction increase in the same ratio as velocity ?

A.—No ; friction does not increase with the velocity at all, if the friction over a given amount of surface be considered ; but it increases as the velocity, if the comparison be made with the time during which the friction acts. Thus the friction of each stroke of a piston is the same, whether it makes 20 strokes in the minute, or 40: in the latter case, however, there are twice the number of strokes made, so that, though the friction per stroke is the same, the friction per minute is doubled. The friction, therefore, of any machine per hour varies as the velocity, though the friction per revolution remains, at all ordinary velocities, the same. Of excessive velocities we have not sufficient experience to enable us to state with confidence whether the same law continues to operate among them.

37. *Q.*—Can you give any approximate statement of the force expended in overcoming friction?

A.—It varies with the nature of the rubbing bodies. The friction of iron sliding upon iron, has generally been taken at about one-tenth of the pressure, when the surfaces are oiled and then wiped again, so that no film of oil is interposed. The friction of iron rubbing upon brass has generally been taken at about one-eleventh of the pressure under the same circumstances; but in machines in actual operation, where a film of some lubricating material is interposed between the rubbing surfaces, it is probably not more than one-third of this amount. Indeed, where unguents are interposed, the friction depends in a great measure upon the nature of the unguent, the viscidity of which may constitute a greater retarding force than the friction; and in watchwork and other fine mechanism, it is therefore necessary to keep the bearing surfaces small, as the resistance by the viscidity of the unguent increases with the extent of the rubbing surface. In some experiments upon the friction of shafts by Mr. G. Rennie, he found that with a pressure of from 1 to 5 cwt. the friction did not exceed $\frac{1}{39}$th of the pressure when tallow was the unguent employed; with soft soap it became $\frac{1}{34}$th. The nature of the unguent, proper for different bearings, appears to depend in a great measure upon the amount of the pressure to which they are subjected,—the hardest unguents being best where the pressure is greatest.

38. *Q.*—What do you understand by a horse power?

A.—An amount of mechanical force that will raise 33,000 lbs. one foot high in a minute. This standard was adopted by Mr. Watt as the average force exerted by the strongest London horses: the object of his investigation being to enable him to determine the relation between the power of a certain size of engine, and the power of a horse; so that when it was desired to supersede the use of horses by the erection of an engine, he might, from the number of horses employed, determine the size of engine that would be suitable for the work.

39. *Q.*—Then, when we talk of an engine of 200-horse power, it is meant that the impelling efficacy is equal to that of 200 horses, each lifting 33,000 lbs. one foot high in a minute?

A.—No, not now; such was the case in Watt's engines, but the capacity of cylinder answerable to a horse power has been increased by most engineers since his time, and the pressure on the piston has been increased also, so that what is now called a 200-horse power engine exerts, almost in every case, a greater power than was exerted in Watt's time, and a horse power has become a mere conventional unit for expressing a certain size of cylinder, without reference to the power exerted.

40. *Q.*—Then each nominal horse power of a modern engine may raise much more than 33,000 lbs. one foot high in a minute?

A.—Yes; some raise 50,000 lbs., others 60,000 lbs. and others 66,000 lbs., one foot high in a minute by each nominal horse power; and therefore no comparison can be made between the performances of different engines, unless the power actually exerted be first discovered.

41. *Q.*—How is this discovery made?

A.—By means of an instrument called the indicator, which consists of a small cylinder, about an inch in diameter, fitted with a piston, which is pressed down by a spring. This piston, by the height to which it rises against the spring, indicates the pressure within the cylinder of the engine; and the number of pounds' pressure on the square inch multiplied by the number of square inches in the area of the cylinder, and by the number of feet travelled through by the piston per minute, gives the amount of impelling force. From this a trifling deduction—about a tenth in the case of large engines is enough—is to be made for friction, &c., and the remainder is the effective moving force which, divided by 33,000 lbs., gives the actual horse power.

42. *Q.*—What quantity of steam is supposed to be consumed by engines per horse power?

A.—About 33 cubic feet per minute, according to

Mr. Watt's rule, or a cubic foot of water raised into steam per hour, which is the proportion assumed by many persons, and which nearly agrees with that adopted by Mr. Watt. Any such rules, however, can only have a very partial application in modern engines, as it is now the common practice to work engines more or less expansively; and, in such cases, less than the quantity of steam formerly necessary, is used.

43. *Q.*—What is meant by working engines expansively?

A.—Adjusting the valves, so that the steam is shut off from the cylinder before the end of the stroke, whereby the residue of the stroke is left to be completed by the expanding steam.

44. *Q.*—And what is the benefit of that practice?

A.—It accomplishes an important saving of steam, or, what is the same thing, of fuel, but it diminishes the power of the engine, while increasing the power of the steam. A larger engine will be required to do the same work; but the work will be done with a smaller consumption of fuel. If, for example, the steam be shut off when only half the stroke is completed, there will only be half the quantity of steam used. But there will be more than half the power exerted; for although the pressure of the steam decreases after the supply entering from the boiler is shut off, yet it imparts, during its expansion, *some* power, and that power, it is clear, is obtained without any expenditure of steam or fuel whatever.

45. *Q.*—Can you give any rule for ascertaining the amount of benefit derivable from expansion?

A.—Divide the length of stroke through which the

steam expands, by the length of stroke performed with full pressure, which call 1 ; the hyperbolic logarithm of the quotient is the increase of efficiency due to expansion. According to this rule it will be found, that if a given quantity of steam, the power of which working at full pressure is represented by 1, be admitted into a cylinder of such a size that its ingress is concluded when one-half the stroke has been performed, its efficacy will be raised by expansion to 1·69 ; if the admission of the steam be stopped at one-third of the stroke, the efficacy will be 2·10 ; at one-fourth, 2·39 ; at one-fifth, 2·61 ; at one-sixth, 2·79 ; at one-seventh, 2·95 ; at one-eighth, 3·08. The expansion, however, cannot be carried beneficially so far as one-eighth, unless the pressure of the steam in the boiler be very considerable, on account of the inconvenient size of cylinder or speed of piston which would require to be adopted, the friction of the engine, and the resistance of vapor in the condenser, which all become relatively greater with a smaller urging force.

46. *Q.*—What is meant by latent heat ?

A.—By latent heat is meant the heat existing in bodies which is not discoverable by the touch or by the thermometer, but which manifests its existence by producing a change of state. Heat is absorbed in the liquefaction of ice, and in the vaporization of water, yet the temperature does not rise during either process, and the heat absorbed is therefore said to become latent. The term is somewhat objectionable, as the effect proper to the absorption of heat has in each case been made visible ; and it would be as reasonable to call hot water latent steam. Latent heat, in the present acceptation of the term, means sensible liquefaction or vaporization; but to produce these changes heat is as necessary as to produce an expansion of the mercury in a thermometer tube, and it is hard to see on what ground heat can be said to be latent when its presence is made manifest by a change of state. It is the *temperature* only that is latent, and latent temperature means sensible something else.

47. *Q.*—But when you talk of the latent heat of steam, what do you mean to express?

A.—I mean to express the heat consumed in accomplishing the vaporization compared with that necessary for producing the temperature. The latent heat of steam is usually reckoned at about 1000 degrees; by which it is meant that there is as much heat in any given weight of steam as would raise its constituent water 1000 degrees if the expansion of the water could be prevented, or as would raise 1000 times that quantity of water one degree. The boiling point is 180 degrees above the freezing point; so that it requires 1180 times as much heat to raise a lb. of water into steam, as to raise 1180 lbs. of water one degree; or it requires about as much heat to raise a pound of boiling water into steam as would raise 5½ lbs. of water from the freezing to the boiling point; 5½ multiplied by 180 being 990, or 1000 nearly.

48. *Q.*—What do you understand by specific heat?

A.—By specific heat, I understand the relative quantities of heat in bodies at the same temperature, just as by specific gravity I understand the relative quantities of matter in bodies of the same bulk. Equal weights of quicksilver and water at the same temperature do not contain the same quantities of heat, any more than equal bulks of those liquids contain the same quantity of matter. The absolute quantity of heat in any body is not known; but the relative heat of bodies at the same temperature, or in other words their specific heats, have been ascertained and arranged in tables, the specific heat of water being taken as unity.

49. *Q.*—What expansion does water undergo in its conversion into steam?

A.—A cubic inch of water makes about a cubic foot of steam of the atmospheric pressure.

50. *Q.*—And how much at a higher pressure?

A.—That depends upon what the pressure is. The higher the pressure, the smaller will be the volume of steam from a given quantity of water; for high pressure steam is just low pressure steam forced into a less space.

51. *Q.*—If this be so, the quantity of heat in a given weight of steam is the same, whether it is high or low pressure steam?

A.—Yes; the heat in steam is a constant quantity, or nearly so, at all pressures, if the steam be saturated with water. Steam, to which an additional quantity of heat has been imparted after leaving the boiler, or as it is called "surcharged steam," comes under a different aw, for the elasticity of such steam may be increased without any addition being made to its weight; but surcharged steam is not employed for working engines, and it may therefore be considered in practice that a pound of steam contains the same quantity of heat at all pressures.

52. *Q.*—But the temperature of steam varies with the pressure?

A.—Yes; the temperature rises with the pressure, but the latent heat becomes less in the same proportion. The latent heat of high pressure steam is less therefore than that of low pressure steam, while the sensible heat is greater, and the two taken together make up the same sum at all temperatures.

53. *Q.*—Is there any benefit arising from the use of high pressure steam in steam engines?

A.—In high pressure, as contrasted with condensing engines, there is always the loss of the vacuum, which will generally amount to 12 or 13 lbs. on the square inch.

54. *Q.*—But in high pressure engines, is there any benefit arising from the use of a very high pressure over a pressure of a moderate account?

A.—Yes, there is an advantage; for, in all high pressure engines, there is a diminution in the power caused by the counteracting pressure of the atmosphere on the educting side of the piston: for the force of the piston in its descent would obviously be greater, if there was a vacuum beneath it; and the counteracting pressure of the atmosphere is relatively less when the steam used is of a very high pressure. It is clear, that if you bring down the pressure of the steam in a high pressure engine to the pressure of the atmosphere, it will not exert any power at all, whatever quantity of steam may be expended, and if the pressure be brought nearly as low as that of the atmosphere, the engine will exert only a very small amount of power; whereas, if a very high pressure be employed, the pressure of the atmosphere will become relatively as small in counteracting the impelling pressure, as the attenuated vapor in the condenser of a condensing engine is in resisting the lower pressure which is there employed.

55. *Q.*—In the case of condensing engines, is there any benefit derivable from the use of steam of a high pressure?

A.—Setting aside loss from friction, and supposing the vacuum to be a perfect one, there would be no benefit arising from the use of steam of a high pressure in condensing engines, for the same weight of steam used without expansion, or with the same measure of expansion, would produce at every pressure the same amount of mechanical power. A piston, with a square foot of area, and a stroke of three feet, with a pressure of one atmosphere, would obviously lift the same weight through the same distance, as a cylinder with half a square foot of area, and a stroke of three feet, with a pressure of two atmospheres. In the one case, we have three cubic feet of steam of the pressure of one atmosphere, and in the other case 1½ cubic feet of the pressure of two atmospheres. But there is the same weight of steam, or the same quantity of heat and water in it, in both cases; so that it appears a given weight of steam would, under such circumstances, produce a definite amount of power, without reference to the pressure. In the case of ordinary engines, however, these conditions do not exactly apply; the vacuum is not a perfect one, and the pressure of the resisting vapor becomes relatively greater as the pressure of the steam is diminished; the friction also becomes greater from the necessity of employing larger cylinders, so that even in the case of condensing engines, there is a benefit arising from the use of steam of a considerable pressure. Expansion cannot be carried beneficially to any great extent, unless the initial pressure be considerable; for if steam of a low pressure were used, the ultimate tension would be reduced to a point so nearly approaching that of the vapor in the condenser, that the difference would not suffice to overcome the friction of the piston; and a loss of power would be occasioned by carrying expansion to such an extent. In some of the Cornish engines, the steam is cut off at one-twelfth of the stroke; but there would be a loss arising from carrying the expansion so far, instead of a gain, unless the pressure of the steam were considerable. It is clear, that in the case of engines which carry expansion very far, a very perfect vacuum in the condenser is more important than it is in other cases. Nothing can be easier than to compute the ultimate pressure of expanded steam, so as to see at what point expansion ceases to be productive of benefit; for as the pressure of expanded steam is inversely as the space occupied, the terminal pressure when the expansion is twelve times, is just one-twelfth of what it was at first, and so on, in all other proportions. The total pressure should be taken as the initial pressure—not the pressure on the safety valve, but that pressure plus the pressure of the atmosphere.

56. *Q.*—Then in high pressure engines working at from 70 to 90 lbs. on the square inch, as in the case of locomotives, the efficiency of a given quantity of water raised into steam may be considered to be about the same as in condensing engines?

A.—Yes, there will not be much difference in the

case supposed: if the pressure of steam in a high pressure engine be 120 lbs., or 105 lbs. above the atmosphere, then the resistance occasioned by the atmosphere will cause a loss of $\frac{1}{8}$th of the power. If the pressure of the steam in a low pressure engine be 16 lbs. on the square inch, or 1 lb. above the atmosphere, and the tension of the vapor in the condenser be equivalent to four inches of mercury, or 2 lbs. of pressure on the square inch, then the resistance occasioned by this rare vapor will also cause a loss of $\frac{1}{8}$th of the power. A high pressure engine, therefore, with a pressure of 105 lbs. above the atmosphere, works with only the same loss from resistance to the piston, as a low pressure engine with a pressure of 1 lb. above the atmosphere; and with these proportions the power produced by a given weight of steam will be the same, whether the engine be high pressure or condensing.

57. *Q.*—Is there any limit to the force of steam?

A.—Yes; there is a limit regulated by the modulus of elasticity of water, which at a temperature of 60° is 22,100 atmospheres. The modulus of elasticity, of any substance, is the measure of its elastic force, and if water be enclosed in a close vessel which it exactly fills, and be then subjected to heat, the expanding force, it is clear, is that of compressed water—for no steam can be produced under such circumstances—and we have this proportion: as the volume of expanded water is to the amount of expansion, so is the modulus of elasticity of water to the elastic force of steam of the same density as water.

58. *Q.*—Have experiments been made to determine the elasticity of steam at different temperatures?

A. — Yes; very careful experiments. The following rule expresses the results obtained by Mr. Southern: — To the given temperature in degrees of Fahrenheit add 51·3 degrees; from the logarithm of the sum, subtract the logarithm of 135·767, which is 2·1327940; multiply the remainder by 5·13, and to the natural number answering to the sum, add the constant fraction ·1, which will give the elastic force in inches of mercury. If the elastic force be known, and it is wanted to determine the corresponding temperature, the rule must be modified thus: — From the elastic force, in inches of mercury, subtract the decimal ·1, divide the logarithm of the remainder by 5·13, and to the quotient add the logarithm 2·1327940; find the natural number answering to the sum, and subtract therefrom the constant 51·3; the remainder will be the temperature sought. The French Academy, and the Franklin Institute, have repeated Mr. Southern's experiments on a larger scale: the results obtained by them are not widely different, and are perhaps nearer the truth, but Mr. Southern's results are generally adopted by engineers, as sufficiently accurate for practical purposes, and those results being identified with engineering practice, it appears expedient to retain them.

59. *Q.*—What law is followed by surcharged steam on the application of heat?

A.—The same as that followed by air, in which the increments in volume are very nearly in the same proportion as the increments in temperature. A volume of air which, at the temperature of 32°, occupies 100 cubic feet, will at 212° fill a space of 137½ cubic feet.

The volume which air or steam out of contact with water, of a given temperature, acquires by being heated to a higher temperature, the pressure remaining the same, may be found by the following rule:—To each of the temperatures before and after expansion, add the constant number 459: divide the greater sum by the less, and multiply the quotient by the volume at the lower temperature; the product will give the expanded volume.

60. *Q.*—If the relative volumes of steam and water are known, is it possible to tell the quantity of water which should be supplied to a boiler, when the quantity of steam expended is specified?

A.—Yes; at the atmospheric pressure, a cubic inch of water has to be supplied to the boiler for every cubic foot of steam abstracted; at other pressures, the relative bulk of water and steam may be determined as follows:—To the temperature of steam in degrees of Fahrenheit, add the constant number 459, multiply the sum by 75·7, and divide the product by the elastic force of the steam in inches of mercury, and the quotient will give the volume required. In practice, however, it is necessary that the feed pump should be able to supply the boiler with a much larger quantity of water than what is indicated by these proportions, from the risk of leaks, priming, or the accidental subsidence of the water to a dangerously low level, which it is necessary as speedily as possible to remedy. It appears expedient that the feed pump should be capable of raising 3½ times the water evaporated by the boiler; it is therefore made about 240th of the capacity of the cylinder for low pressure engines, supposing the cylinder to be double acting, and the pump single acting. In high pressure engines, however, the capacity of the pump should be greater, in proportion to the pressure of the steam.

61. *Q.*—How do you estimate the quantity of watei requisite for condensation?

A.—Mr. Watt found that the most beneficial temperature of the hot well of his engines was 100 degrees. If, therefore, the temperature of the steam be 212°, and the latent heat 1,000°, then 1,212° may be taken to represent the heat contained in the steam, or 1,112° if we deduct the temperature of the hot well. If the temperature of the injection water be 50°, then 50 degrees of cold are available for the abstraction of heat; and as the total quantity of heat to be abstracted, is that requisite to raise the quantity of water in the steam 1,112 degrees, or 1,112 times that quantity one degree, it would raise one-fiftieth of this, or 22·24 times the quantity of water in the steam, 50 degrees. A cubic inch of water therefore raised into steam, will require 22·24 cubic inches of water at 50 degrees for its condensation, and will form therewith 23·24 cubic inches of hot water at 100 degrees. Mr. Watt's practice was to allow about a wine pint (28·9 cubic inches) of injection water, for every cubic inch of water evaporated from the boiler. The usual capacity for the cold water pump is $\frac{1}{48}$th of the capacity of the cylinder, which allows some water to run to waste.

62. *Q.*—Is not a good vacuum in an engine conducive to increased power?

A.—It is.

63. *Q.*—And is not the vacuum good in the pro-

portion in which the temperature is low, supposing there to be no air leaks?

A.—Yes.

64. *Q.*—Then how could Mr. Watt find a temperature of 100° in the water drawn from the condenser, to be more beneficial than a temperature of 70° or 80°, supposing there to be an abundant supply of cold water?

A.—Because the superior vacuum due to a temperature of 70° or 80° involves the admission of so much cold water into the condenser, which has afterwards to be pumped out in opposition to the pressure of the atmosphere, that the gain in the vacuum does not equal the loss of power occasioned by the additional load upon the pump; and there is therefore a clear loss by the reduction of the temperature below 100°, if such reduction be caused by the admission of an additional quantity of water. If the reduction of temperature, however, be caused by the use of colder water, there is a gain produced by it, though the gain will within certain limits be greater, if advantage be taken of the lowness of the temperature to diminish the quantity of injection.

65. *Q.*—Cannot the condensation of the steam be accomplished by any other means, than by the admission of cold water into the condenser?

A.—It may be accomplished by the method of external cold, as it is called, which consists in the application of a large number of thin metallic surfaces to the condenser, on the one side of which the steam circulates, while on the other side there is a constant current of cold water, and the steam is condensed by coming into contact with the cold surfaces, without mingling with the water used for the purpose of refrigeration. The first kind of condenser employed by Mr. Watt was constructed after this fashion, but he found it in practice to be inconvenient from its size, and to become furred up or incrusted when the water was bad, whereby the conducting power of the metal was impaired. He therefore reverted to the use of the jet of cold water, as being upon the whole preferable. The jet entered the condenser instead of the cylinder as was the previous practice, and this method is now the one in common use. Some few years ago, a good number of steam vessels were fitted with Hall's condensers, which operated on the principle of external cold, and which consisted of a fagot of small copper tubes surrounded by water; but the use of those condensers has not been persisted in, and most of the vessels fitted with them have returned to the ordinary plan.

66. *Q.*—How much water will a pound of coal raise into steam?

A.—From 6 to 8 lbs. of water in the generality of land boilers of medium quality, the difference depending on the kind of boiler, the kind of coal, and other circumstances. Mr. Watt reckoned his boilers as capable of evaporating 10·08 cubic feet of water with a bushel or 84 lbs. of coal, which is equivalent to 7½ lbs. of water evaporated by one pound of coal, and this may be taken as the performance of common land boilers at the present time. In some of the Cornish boilers, however, a pound of coal raises 10 lbs. of water into steam, or a cwt. of coal evaporates about 19 cubic feet of water. The quantity of fuel burned on each square foot of fire grate per hour, varies very much in different boilers:

in wagon boilers it is from 10 to 13 lbs.; in Cornish boilers from 3½ to 4 lbs.; and in locomotive boilers from 80 to 100 lbs. The number of square feet of surface required to evaporate a cubic foot of water per hour is about 70 square feet in Cornish boilers, 9 to 11 square feet in land and marine boilers, and about 6 square feet in locomotive boilers. The number of square feet of heating surface per square foot of fire grate, is from 13 to 15 square feet in wagon boilers; about 40 square feet in Cornish boilers; and from 50 to 70 square feet in locomotive boilers.

67. *Q.*—But what is the heating surface per horse power?

A.—About 9 square feet per horse power is the usual proportion in wagon boilers, reckoning the total surface as effective surface, if the boilers be of a considerable size; but in the case of small boilers the proportion is larger. The total heating surface of a two-horse power wagon boiler is, according to Boulton and Watt's proportions, 30 square feet or 15 ft. per horse power; whereas, in the case of a 45-horse power boiler the total heating surface is 438 square feet or 9·6 ft. per horse power. The capacity of steam room is 8¾ cubic feet per horse power, in the two-horse power boiler, and 5¾ cubic feet in the 20-horse power boiler; and in the larger class of boilers, such as those suitable for 30 and 45 horse power engines, the capacity of the steam room does not fall below this amount, and indeed is nearer 6 than 5¾ cubic feet per horse power. The content of water is 18½ cubic feet per horse power in the two-horse power boiler, and 15 cubic feet per horse power in the 20-horse power boiler. In marine boilers about the same proportions obtain in most particulars. The original boilers of the "Great Western" steamer, by Messrs. Maudslay, were proportioned with about half a square foot of fire grate per horse power, and ten square feet of flue and furnace surface, reckoning the total amount as effective; but in the boilers of the "Retribution," by the same makers, a somewhat smaller proportion of heating surface was adopted. Boulton and Watt have found that in their marine flue boilers 9 square feet of flue and furnace surface are requisite to boil off a cubic foot of water per hour, which is the proportion that obtains in their land boilers; but inasmuch as in modern engines the nominal considerably exceeds the actual power, they allow 11 square feet of heating surface per nominal horse power in their marine boilers, and they reckon as effective heating surface, the tops of the flues, and the whole of the sides of the flues, but not the bottoms. They have been in the habit of allowing for the capacity of the steam space in marine boilers 16 times the content of the cylinder; but as there are two cylinders, this is equivalent to 8 times the content of both cylinders, which is the proportion commonly followed in land engines, and which agrees very nearly with the proportion of between 5 and 6 cubic feet of steam room per horse power. Taking, for example, an engine with 23 inches diameter of cylinder and 4 feet stroke—which will be 18·4 horse power — the area of the cylinder will be 415·476 square inches, which multiplied by 48, the number of inches in the stroke, will give 19942·848 for the capacity of the cylinder in cubic inches; 8 times this is 159542·784 cubic inches, or 92·3 cubic feet; 92·3 divided by 18·4 is rather more than 5 cubic feet per horse power. There is less necessity, however, that the steam space should be large when the flow of steam from the boiler is very uniform, as it will be where there are two engines attached to the boiler at right angles with one another, or where the engines work at a great speed, as in the case of locomotive engines. A high steam chest too, by rendering boiling over into the steam pipes, or priming as it is called, more difficult, obviates the necessity for so large a steam space; and the use of steam of a high pressure, worked expansively, has the same operation; so that in modern marine boilers, of the tubular construction, where the whole of these modifying circumstances exist, there is no necessity for so large a proportion of steam room as 5 or 6 cubic feet per horse power, and about half that amount more nearly represents the general practice. In many of the marine tubular boilers, however, the steam space is made too small; in some cases only $1\frac{3}{4}$ cubic feet per horse power; but the operation of boilers thus proportioned is not satisfactory. The proportion of 11 square feet of heating surface per nominal horse power, reckoning all the surface as effective, half a square foot of fire grate, and 3 cubic feet of steam room, seems to answer very well for tubular boilers, where the steam chest rises above the boiler, and the engines work expansively through one-third or one-fourth of the stroke. Boulton and Watt allow 0·64 of a square foot per nominal horse power of grate bars in their marine boilers, and a good effect arises from this proportion; but sometimes so large an area of fire grate cannot be conveniently got, and the proportion of half a square foot per horse power seems to answer very well in engines working with some expansion, and is now very widely adopted. With this allowance, there will be about 22 square feet of heating surface per square foot of fire grate, and if the consumption of fuel be taken at 6 lbs. per nominal horse power per hour, there will be 12 lbs. of coal consumed per hour on each square foot of grate. The furnaces should not be more than 6 ft. long, as, if much longer than this, it is impossible to fire them effectually.

68. *Q.*—Can the consumption be made as small as 6 lbs. per nominal horse power per hour?

A.—It can by working expansively, and in some engines the consumption per nominal horse power per hour does not exceed this amount.

69. *Q.*—Can the consumption be made as low as 6 lbs. per actual horse power per hour without resorting to expansion?

A.—An actual horse power, or 33,000 lbs. raised one foot high in the minute, is represented by the evaporation of a cubic foot of water in the hour; and if a cubic foot of water has to be evaporated by 6 lbs. of coal, then 1 lb. of coal must be capable of evaporating 10½ lbs. of water, which is more than the usual proportion, and engines, therefore, with such a rate of consumption, may be set down as deriving a part of their economy from expansion.

70. *Q.*—Would it not be beneficial to introduce the Cornish boilers into steam vessels, since they operate with a superior economy in fuel?

A.—No; it would not. The superior economy of the Cornish boilers is not derived from any peculiarity of form or arrangement, but from the immense extension of heating surface, which, though productive of economy in fuel, is attended with an increased expense of boiler, while the space occupied is such as in the case of a steam vessel would be quite inadmissible. If a marine boiler be made with the same quantity of surface as a Cornish boiler, the marine boiler is found to be the more economical of the two.

71. *Q.*—Is not the combustion in the furnaces of the Cornish boilers very slow?

A.—Yes, very slow; and there is in consequence very little smoke evolved. It is an important part of smoke burning to make the combustion slow; and the chief instigator of smoke, is an insufficient size of furnace. The coal used in Cornwall is Welsh coal, which evolves but little smoke, and is therefore more favorable for the success of a smokeless furnace; but in the manufacturing districts, where the coal is more bituminous, it is found that smoke may be almost wholly prevented by careful firing and a large capacity of furnace.

72. *Q.*—What is the nature of combustion?

A.—Combustion is nothing more than an energetic chemical combination; or, in other words, it is the mutual neutralization of opposing electricities. When coal is brought to a high temperature it acquires a strong affinity for oxygen, and combination with oxygen will produce more than sufficient heat to maintain the original temperature.

73. *Q.*—Does air consist of oxygen?

A.—Air consists of oxygen and nitrogen mixed together, in the proportion of 3½ lbs. of nitrogen to 1 lb. of oxygen. Every pound of coal requires about two pounds of oxygen for its saturation, and therefore for every pound of coal burned seven pounds of nitrogen must pass through the fire, supposing all the oxygen to enter into combination. In practice, however, this perfection of combination does not exist: from one-third to one-half of the oxygen will pass through the fire without entering into combination; so that from 12 to 14 lbs. of air are required for every pound of coal burned. 12 lbs. of air are about 160 cubic feet, and 200 cubic feet may be taken as the quantity of air required for the combustion of a pound of coal in practice.

74. *Q.*—Had Mr. Watt any method of consuming smoke?

A.—He tried various methods, but eventually fixed upon the method of coking the coal on a dead plate at the furnace door, before pushing it into the fire. That method is perfectly effectual where the combustion is so slow that the requisite time for coking is allowed, and it is much preferable to any of the methods of admitting air at the bridge or elsewhere, to accomplish the combustion of the inflammable parts of the smoke. But the best arrangement for smoke burning is, it appears to me, that of the revolving grate, which feeds the fire at the same time by a self-acting mechanism. In this plan, the fire grate is made like a round table capable of turning upon a centre; a shower of coal is precipitated upon the grate through a slit in the boiler near the furnace mouth, and the smoke evolved from the coal dropped at the front part of the fire is consumed by passing over the incandescent fuel at the back part, from which all the smoke must have been expelled in the revolution of the grate before it can have reached that position.

75. *Q.*—Is a furnace of this kind applicable to a steam vessel?

A.—I see nothing to prevent its application. The arrangement of the boiler should perhaps be changed, to facilitate its application, and I should prefer to combine its use with the employment of vertical tubes, which are found to be more efficacious in practice than horizontal tubes; but this innovation is not indispensable. The introduction of any effectual automatic contrivance for feeding the fire in steam vessels, would bring about an important economy, at the same time that it would give the assurance of the work being better done. It is very difficult to fire furnaces by hand effectually at sea, especially in rough weather and in tropical climates; whereas machinery would be unaffected by any such impediments, and would perform with little expense the work of many men.

76. *Q.*—The introduction of some mechanical method of feeding the fire with coals would enable a double tier of furnaces to be adopted in steam vessels without inconvenience?

A.—Yes, it would have at least that tendency; and as the space available for area of grate is limited in a steam vessel by the width of the vessel, it would be a great convenience if a double tier of furnaces could be employed without a diminished effect. It appears to me, however, that the objection would still remain of the steam raised by the lower furnace being cooled and deadened by the air entering the ash-pit of the upper fire, for it would strike upon the metal of the ash-pit bottom; but this surface could be covered with fire-brick to diminish the refrigeratory effect.

77. *Q.*—What method of firing furnaces is the best?

A.—The coals should be broken up into small pieces, and sprinkled thinly and evenly over the fire a little at a time. The thickness of the stratum of coal upon the grate should depend upon the intensity of the draft: in ordinary land or marine boilers it should be thin, whereas in locomotive boilers it requires to be much thicker. If the stratum of coal be thick while the draft is sluggish, the carbonic acid resulting from combustion combines with an additional atom of carbon in passing through the fire, and is converted into carbonic oxide, which may be defined to be invisible smoke, as it carries off a portion of the fuel: if, on the contrary, the stratum of coal be thin while the draft is very rapid, an injurious refrigeration is occasioned by the excess of air passing through the furnace. The fire should always be spread of uniform thickness over the bars of the grate, and should be without any holes or uncovered places, which greatly diminish the effect of the fuel by the refrigeratory action of the stream of cold air which enters thereby.

78. *Q.*—What is the method of consuming smoke pursued in the manufacturing districts?

A.—In Manchester, where some stringent regulations for the prevention of smoke have latterly been enacted, it is found that the readiest way of burning the smoke is to have a very large proportion of furnace room, whereby slow combustion may be carried on, and the operation of the dead plate, prescribed by Mr. Watt, may be made more effectual. In some cases, too, a favorable result is arrived at by raising a ridge of coal across the furnace lying against the bridge, and of the same height: this ridge speedily becomes a mass of coke, which promotes the combustion of the smoke passing over it.

78. *Q.*—Is the method of admitting a stream of air into the flues regarded favorably?

A.—No; it is found to be productive of injury to the boiler by the violent alternations of temperature it occasions, as at some times cold air impinges on the iron of the boiler, and at other times flame, just as there happens to be smoke or no smoke emitted by the furnace. Boilers, therefore, operating upon this principle, speedily become leaky, and are much worn by oxidation, so that, if the pressure is considerable, they are liable to explode. It is very difficult to apportion the quantity of air admitted, to the varying wants of the fire, and as air may at some times be rushing in when there is no smoke to consume, a loss of heat, and an increased consumption of fuel may be the result of the arrangement; and, indeed, such is the result in practice, though a carefully performed experiment usually demonstrates a saving of 10 or 12 per cent.

80. *Q.*—What other plans have been contrived for obviating the nuisance of smoke?

A.—They are too various for enumeration, but most of them either operate upon the principle of admitting air into the flues to accomplish the combustion of the uninflammable parts of the smoke, or seek to attain the same object by passing the smoke over or through the fire, or other incandescent materials. Some of the plans, indeed, profess to burn the inflammable gases as they are evolved from the coal, without permitting the admixture of any of the uninflammable products of combustion which enter into the composition of smoke; but this object has been very imperfectly fulfilled in any of the contrivances yet brought under the notice of the public, and as the revolving grate, without aspiring to a theoretic perfection, seems to achieve the combustion of the smoke when the fire is not urged with vehemence, and as it saves in addition the labor of firing, it seems to be entitled to a preference in practice. Jucke's furnace, which consists of an arrangement of the fire bars in the form of an endless chain, whereby the coal is gradually carried forward as it burns, and the ashes and clinkers are precipitated into the ash-pit at the extremity of the furnace where the chain turns down over the extreme carrying roller, is also well spoken of; but it is more expensive to construct than the revolving grate, and more difficult to keep in repair. The patent-right of the revolving grate expired some years ago, which may, perhaps, be in some cases a further suasory to its adoption.

81. *Q.*—You have stated that an *actual* horse power is a force capable of raising 33,000 lbs. one foot high in the minute, or a dynamical effort such as is produced by an expenditure of 33 cubic feet of steam per minute, or the evaporation of a cubic foot of water per hour, but you have given no definition of the value of a *nominal* horse power. How do you ascertain the power of an engine in nominal horse power?

A.—The nominal power of an engine may be ascertained by the following rule: multiply the square of the diameter of the cylinder in inches by the velocity of the piston in feet per minute, and divide the product by 6000 ; the quotient is the number of nominal horse power. In using this rule, however, it is necessary to adopt the speed of piston described by Mr. Watt, which varies with the length of the stroke. The speed of piston with a two feet stroke is, according to his system, 160 per minute ; with a 2 ft. 6 in. stroke, 170 ; 3 ft., 180 ; 3 ft. 6 in., 189 ; 4 ft., 200 ; 5 ft., 215 ; 6 ft., 228 ; 7 ft., 245 ; 8 ft., 256 ft.

82. *Q.*—By ascertaining the ratio in which the velocity of the piston increases with the length of the stroke, could not the element of velocity be cast out altogether, and the nominal power be determined by a reference merely to the dimensions of the cylinder ?

A.—Yes, and this for most purposes is the most convenient method of procedure: multiply the square of the diameter of the cylinder in inches by the cube root of the stroke in feet, and divide the product by 47 ; the quotient is the number of nominal horse power of the engine. This rule supposes a uniform effective pressure upon the piston of 7 lbs. per square inch. Mr. Watt estimated the effective pressure upon the piston of his 4-horse power engines at 6·8 lbs. per square inch, and the pressure increased slightly with the power, and became 6·94 lbs. per square inch in engines of 100-horse power; but it appears to be more convenient to take a uniform pressure of 7 lbs. for all powers. Small engines, indeed, are somewhat less effective in proportion than large ones, but the difference can be made up by slightly increasing the pressure in the boiler; and small boilers will bear such an increase without inconvenience.

83. *Q.*—Can nominal be transformed into actual horse power?

A.—No, that is not possible in the case of common condensing engines; the actual power exerted by an engine cannot be deduced from its nominal power, neither can the nominal power be deduced from the power actually exerted, nor from any thing else than the dimensions of the cylinder. The actual horse power is a dynamical unit, and the nominal horse power is a measure of capacity of the cylinder, which are obviously incomparable things.

84. *Q.*—That is, the nominal horse power expresses the size of an engine, and the actual horse power the number of times 33,000 lbs. it will lift one foot high in a minute?

A.—Precisely; and to find the number of times 33,000 lbs., or 528 cubic feet of water, it will raise one foot high in a minute—or, in other words, the actual power—you first find the pressure in the cylinder by means of the indicator, from which you deduct a pound and a half of pressure for friction, the loss of power in working the air pump, &c.; multiply the area of the piston in square inches by this residual pressure, and by the motion of the piston, in feet per minute, and divide by 33,000; the quotient is the actual number of horses power. The same result is attained by squaring the diameter of the cylinder, multiplying by the pressure per square inch, as shown by the indicator, less a pound and a half, and by the motion of the piston, in feet, and dividing by 42,017. The quantity thus arrived at will, in the case of nearly all modern engines, be very different from that obtained by multiplying the square of the diameter of the cylinder by the cube root of the stroke, and dividing by 47, which expresses the nominal power; and the actual and nominal power must by no means be confounded, as they are totally different things.

85. *Q.*—What is meant by the duty of an engine?

A.—The work done in relation to the fuel consumed.

86. *Q.*—And how is the duty ascertained?

A.—In ordinary mill or marine engines it can only be ascertained by the indicator, as the load upon such engines is variable, and cannot readily be determined; but in the case of engines for pumping water, where the load is constant, the number of strokes performed by the engine will represent the work done, and the amount of work done by a given quantity of coal represents the duty. In Cornwall the duty of an engine is expressed by the number of millions of pounds raised one foot high by a bushel, or 94 lbs. of Welsh coal. A bushel of Newcastle coal will only weigh 84 lbs.; and in comparing the duty of a Cornish engine with the performance of an engine in some locality where a different kind of coal is used, it is necessary to pay regard to such variations.

87. *Q.*—Can you tell the duty of an engine when you know its consumption of coal per horse power per hour?

A.—Yes, if the power given be the actual, and not the nominal power. Divide 166·32 by the number of pounds of coal consumed per actual horse power per hour; the quotient is the duty in millions of pounds. If you already have the duty in millions of pounds, and wish to know the equivalent consumption in pounds per actual horse power per hour, divide 166·32 by the duty in millions of pounds; the quotient is the consumption per actual horse power per hour. The duty of a locomotive engine is expressed by the weight of coke it consumes in transporting a ton through the distance of one mile upon a railway; but this is a very imperfect method of representing the duty, as the tractive efficacy of a pound of coke becomes less as the speed of the locomotive becomes greater, and the law of variation is not accurately known.

88. *Q.*—What amount of power is generated in good engines of the ordinary kind by a given weight of coal?

A.—The duty of different kinds of engines varies very much, and there are also great differences in the performance of different engines of the same class. In ordinary rotative condensing engines of good construction, 10 lbs. of coal per nominal horse power per hour is a common consumption; but such engines exert nearly twice their nominal power, so that the consumption per actual horse power per hour may be taken at from 5 to 6 lbs. Engines working very expansively, however, attain an economy much superior to this. The average duty of the pumping engines in Cornwall is about 60,000,000 lbs. raised 1 ft. high by a bushel of Welsh coals, which weighs 94 lbs. This is equivalent to a consumption of 3·1 lbs. of coal per actual horse power per hour; but some engines reach a duty of above 100,000,000, or 1·74 lbs. of coal per actual horse power per hour. Locomotives consume from 8 to 10 lbs. of coke in evaporating a cubic foot of water, and the evaporation of a cubic foot of water per hour may be set down as representing an actual horse power in locomotives as well as in condensing engines. Measuring the consumption of fuel by the number of tons a locomotive will draw through a given distance, it appears that passenger engines consume from $\frac{1}{2}$ to $\frac{3}{4}$ lb. of coke per ton per mile, and the goods engines from $\frac{1}{4}$ to $\frac{1}{2}$ lb. When the locomotive is worked expansively, there is of course a less consumption of water and fuel per horse power, or per ton per mile, than when the full pressure is used throughout the stroke, and most locomotives now operate with as much expansion as can be conveniently given by the slide valves.

89. *Q.*—But some kinds of coal are more effective than other kinds, and there will be a difference of effect from this cause?

A.—Yes. In a boiler so proportioned, that a pound of the best Welsh coal will evaporate 9·493 lbs. of water; a pound of anthracite will evaporate 9·014 lbs. of water; a pound of the best small Newcastle 8·524 lbs.; a pound of coke from gas works 7·908 lbs.; a pound of Welsh and Newcastle, of medium quality, mixed half and half, 7·897 lbs.; a pound of Derbyshire 6·772 lbs.; and a pound of Blyth Main, Northumberland, 6·6 lbs. of water. There thus appears to be a difference of at least one-third in the efficacy of the different coals of commerce in raising steam. Coal and coke may be reckoned equal to one another in evaporative power. A pound of wood in its ordinary state evaporates about 4·72 lbs. of water, and a pound of turf 5·45 lbs. A pound of wood charcoal, however, is said to evaporate 13·37 lbs., and a pound of turf charcoal 11·63 lbs. The number of pounds of air consumed by a pound of any combustible is very nearly the same as the number of pounds of water it is capable of evaporating, though twice that quantity of air must pass through the fire. A pound of wood in its ordinary state requires 4·47 lbs. of air for its combustion, and a pound of turf 4·6 lbs. If one pound of coal evaporate 9½ lbs. of water, then a cubic foot of water will be evaporated by 6·58 lbs. of coal. 6½ lbs. of water evaporated by a pound of coal is equivalent to a cubic foot of water evaporated by 9·61 lbs. of coal; and 4½ lbs. of water evaporated by a pound is equivalent to a cubic foot of water evaporated by 13·88 lbs. of coal. In such of the Cornish boilers as evaporate 10 lbs. of water with a pound of coal, a cubic foot of water will be evaporated by 6¼ lbs. of coal; but the usual performance of the best Cornish boilers is about 9 lbs. of water evaporated by a pound of coal of the best quality.

90. *Q.*—By what process do you ascertain the dimensions of the chimney?

A.—By a reference to the volume of air it is necessary in a given time to supply to the burning fuel, and to the velocity of motion produced by the rarefaction in the chimney; for the area of the chimney requires to be such, that with the velocity due to that rarefaction, the quantity of air requisite for the combustion of the fuel shall pass through the furnace in the specified time. Thus if 200 cubic feet of air of the atmospheric density are required for the combustion of a pound of coal, and 10 lbs. of coal per horse power per hour are consumed by an engine, then 2000 cubic feet of air must be supplied to the furnace per horse per hour; and the area of the chimney must be such, as to deliver this quantity at the increased bulk due to the high temperature of the chimney when moving with the velocity the rarefaction within the chimney occasions, and which is usually such as to support a column of half an inch of water. The velocity with which a denser fluid flows into a rarer one is equal to the velocity a heavy body acquires in falling through a height equal to the difference of altitude of two columns of the heavier fluid such as will produce the respective pressures; and, therefore, when the difference of pressure or amount of rarefaction in the chimney is known, it is easy to tell the velocity of motion which ought to be produced by it. In practice, however, these theoretical results are not to be trusted, until they have received such modifications as will make them representative of the practice of the most experienced constructors. Boulton and Watt's rule for the dimensions of the chimney of a land engine is as follows:—multiply the number of pounds of coal consumed under the boiler per hour by 12, and divide the product by the square root of the height of the chimney in feet; the quotient is the area of the chimney in square inches in the smallest part. A factory chimney suitable for a 20-horse boiler is commonly made about 20 in. square inside, and 80 ft. high, and these dimensions are those which answer to a consumption of 15 lbs. of coal per horse power per hour, which is a very common consumption in factory engines. If 15 lbs. of coal be consumed per horse power per hour, the total consumption per hour in a 20-horse boiler will be 300 lbs., and 300 multiplied by 12=3600, and divided by 9 (the square root of the height)=400, which is the area of the chimney in square inches. It will not answer well to increase the height of a chimney of this area to more than 40 or 50 yards, without also increasing the area, nor will it be of utility to increase the area much without also increasing the height. The quantity of coal consumed per hour in pounds, multiplied by 5, and divided by the square root of the height of the chimney, is the proper collective area of the openings between the bars of the grate for the admission of air to the fire.

91. *Q.*—Is this rule for the dimensions of the chimney also applicable to steam vessels?

A.—In steam vessels Boulton and Watt allow 8½ square inches of area of chimney per horse power, and in marine flue boilers they allow 18 square inches of sectional area of flue per horse power; but this proportion appears to be about one-third greater than what is allowed by many other makers, whose boilers, however, are scarcely so conspicuous for an abundant supply of steam. The sectional area of the flue in square inches is what is termed the *calorimeter* of the boiler, and the calorimeter divided by the length of the flue in feet is what is termed the *vent*. In marine flue boilers of good construction the vent varies between the limits of 21 and 25, according to the size of the boiler and other circumstances—the largest boilers having generally the largest vents; and the calorimeter divided by the vent will give the length of the flue in feet. The collective area for the escape of the smoke and flame over the furnace bridges in marine boilers is 19 square inches per horse power, according to Boulton and Watt's proportion.

92. *Q.*—Are these proportions applicable to tubular and wagon boilers?

A.—No. In wagon and tubular boilers very different proportions prevail, yet the proportions of every kind of boiler are determinable on the same general principle. In wagon boilers the proportion of the perimeter of the flue which is effective as heating surface, is to the total perimeter as 1 to 3, or, in some cases, as 1 to 2·5; and with any given area of flue, therefore, the length of the flue must be from 3 to 2·5 times greater than would be necessary if the total surface were effective. If then the vent be the calorimeter divided by the length, and the length be made 3 or 2·5 times greater, the vent must become 3 or 2·5 times less; and in wagon boilers accordingly the vent varies from 8 to 11 instead of from 21 to 25, as in the case of marine flue boilers. In Boulton and Watt's 45-horse wagon boiler the area of flue is 18 square inches per horse power; but the area per horse power increases very rapidly as the size of the boiler becomes less, and amounts to about 80 square inches per horse power in a boiler of two-horse power. Some such increase is obviously inevitable if a similar form of flue be retained in the larger and smaller powers, and at the same time the elongation of the flue in the same proportion as the increase of any other dimension is prevented; but in the smaller class of wagon boilers the consideration of facility of cleaning the flues is also operative in inducing a large proportion of sectional area. Boulton and Watt's 2-horse power wagon boiler has 30 square feet of surface, and the flue is 18 in. high above the level of the boiler bottom, by 9 in. wide; while their 12-horse wagon boiler has 118 square feet of heating surface, and the dimensions of the flue similarly measured are 36 in. by 13 in. The width of the smaller flue, if similarly proportioned to the larger one, would be 6½ in., instead of 9 in.; and, by assuming this dimension, we should have the same proportion of sectional area per square foot of heating surface in both boilers. The length of flue in the 2-horse boiler is 19·5 ft., and in the 12-horse boiler 39 ft., so that the length and height of the flue are increased in the same proportion.

93. *Q.*—Will you extend your illustrations to the case of a marine boiler?

A.—The Nile steamer, with engines of 110-horse power, by Boulton and Watt, is supplied with steam by two boilers, which are, therefore, of 55-horse power each. The height of the flue winding within the boiler is 60 in., and its mean width 16½ in., making a sectional area or calorimeter of 990 square inches, or 18 square inches per horse power of the boiler. The length of the flue is 39 ft., making the vent 25, which is the vent proper for large boilers. In the Dee and Solway steamers, by Scott and Sinclair, the calorimeter is only 9·72 square inches per horse power; in the Eagle, by Caird, 11·9; in the Thames and Medway, by Maudslay, 11·34, and in a great number of other cases it does not rise above twelve square inches per horse power; but the engines of most of these vessels are intended to operate to a certain extent expansively, and the boilers are less powerful in evaporating efficacy on that account.

94. *Q.*—Then the chief difference in the proportions

established by Boulton and Watt, and those followed by the other manufacturers you have mentioned is, that Boulton and Watt set a more powerful boiler to do the same work ?

A.—That is the main difference. The proportion which one part of the boiler bears to another part is very similar in the cases cited, but the proportion of boiler relatively to the size of the engine varies very materially. Thus the calorimeter *of each boiler* of the Dee and Solway is 1,296 square inches ; of the Eagle, 1,548 square inches ; and of the Thames and Medway, 1,134 square inches ; and the length of flue is 57, 60, and 52 ft. in the boilers respectively, which makes the respective vents 22½, 25, and 21 in. Taking then the boiler of the Eagle for comparison with the boiler of the Nile, as it has the same vent, it will be seen that the proportions of the two are almost identical, for 990 is to 1,548 as 39 is to 60, nearly; but Messrs. Boulton and Watt would not have set a boiler like that of the Eagle to do so much work.

95. *Q.*—Then the evaporating power of the boiler varies as the sectional area of the flue ?

A.—The evaporating power varies as the square root of the area of the flue, if the length of the flue remain the same ; but it varies as the area simply, if the length of the flue be increased in the same proportion as its other dimensions. The evaporating power of a boiler is referable to the amount of its heating surface, and the amount of heating surface in any flue or tube is proportional to the product of the length of the tube and the square root of its sectional area, multiplied by a certain quantity that is constant for each particular form. But in similar tubes the length is proportional to the square root of the sectional area; therefore, in similar tubes, the amount of heating surface is proportional to the sectional area. On this area also depends the quantity of hot air passing through the flue, supposing the intensity of the draft to remain unaffected, and the quantity of hot air or smoke passing through the flue should vary in the same ratio as the quantity of surface.

96. *Q.*—A boiler, therefore, to exert four times the power should have four times the extent of heating surface, and four times the sectional area of flue for the transmission of the smoke ?

A.—Yes ; and if the same form of flue is to be retained it should be of twice the diameter and twice the length ; or twice the height and width if rectangular, and twice the length. As then the diameter or square root of the area increases in the same ratio as the length, the square root of the area divided by the length ought to be a constant quantity in each type of boiler, in order that the same proportions of flue may be retained ; and in wagon boilers without an internal flue, the height in inches of the flue encircling the boiler divided by the length of the flue in feet will be 1 very nearly. Instead of the square root of the area, the effective perimeter, or outline of that part of the cross section of the flue which is effective in generating steam, may be taken ; and the effective perimeter divided by the length ought to be a constant quantity in similar forms of flue, and with the same velocity of draft, whatever the size of the flue may be. It is clear, that with any given area of flue, to increase the perimeter by adopting a different shape, is to diminish the length of the flue; and, if the extent of the perimeter be diminished, the length of the flue must at the same time be increased, else it will be impossible to obtain the necessary amount of heating surface. In Boulton and Watt's wagon boilers the sectional area of the flue in square inches per square foot of heating surface, is 5·4 in the two-horse boiler; in the three-horse it is 4·74; in the four-horse, 4·35; six-horse, 3·75; eight-horse, 4·33; ten-horse, 3·96; twelve-horse, 3·63; eighteen horse, 3·17; thirty-horse, 2·52, and in the forty-five horse boiler, 2·05 square inches. Taking the amount of heating surface in the forty-five horse boiler at 9 square feet per horse power, we obtain 18 square inches of sectional area of flue per horse power, which is also Boulton and Watt's proportion of sectional area for marine boilers with internal flues.

97. *Q.*—If to increase the perimeter of a flue is virtually to diminish the length, then a tubular boiler where the perimeter is in effect greatly extended ought to have but a short length of tube?

A.—The flue of the Nile, if reduced to the cylindrical form, would be 35½ in. in diameter to have the same area; but it would then require to be made 47¾ ft. long, to have the same amount of heating surface. Supposing that with these proportions the heat is sufficiently extracted from the smoke, then every tube of a tubular boiler in which the same draft existed ought to have very nearly the same proportions, so that a tube 3 in. in diameter ought to be about 4 ft. long, supposing the conducting power of the metallic surface through which the heat is transmitted, to be in each case identical. But the metal of small tubes being thinner than that of flues must conduct better, and a tube 3 in. in diameter should therefore be less than 4 ft. long, provided the draft remains such as is due to an area of 18 square inches per horse power. If the thinness of the metal attainable by the tubular form be supposed to increase the efficacy of the heating surface in the same proportion as the increase of surface due to the rectangular form, the length of a tube 3 in. diameter ought to be 3 ft. 3 in., and it would be of no service to extend its length beyond this point, supposing the flue boiler to be properly proportioned, as by the time the hot air had traversed a length of 3 ft. 3 in. of tube, the heat of the air would have been as thoroughly extracted as in ordinary boilers appears to be beneficial. The tubes of tubular boilers, however, are usually about 6 ft. 6 in. long; but to make this excess of length influential in generating steam, the draft has to be made nearly twice greater than in flue boilers having a sectional area of 18 square inches per horse power; or, in other words, the sectional area of tubular boilers must not much exceed 9 square inches per horse power when the tubes are of the length stated. The smaller the tubes are, the shorter they should be made, or the less the sectional area ought to be; and with a sectional area of 10 square inches per horse power there will be no advantage in making the length of the tube more than from 26 to 32 times its diameter, which will make the tube from 6 ft. 6 in. to 8 ft. long when the diameter is three inches, and give from 7·4 to 8 square feet of heating surface of tubes per horse power. If the sectional area per horse power be increased, the length of tube should be diminished in the same proportion; for the velocity of the draft varies with the sectional area of tube per horse power, and on the velocity of draft the length of the tube ought to depend. In locomotive boilers where the velocity of draft is very great, long tubes are employed; but it is preferable to have the tubes of moderate length and a draft of a moderate intensity, as in maintaining a fierce draft by any process, there is a considerable expenditure of power. If, however, with the view of making the draft very slow, a proportion of sectional area approaching that of flue boilers be provided, the result will not be satisfactory, as the smoke will all pass through a few of the tubes, leaving the rest inoperative; though this defect may be in a great measure corrected by partially closing up the ends of the tubes, or even by partially closing the damper. The length of tube multiplied by the diameter, and divided by the area, is a constant quantity both in flue and tubular boilers, or at least nearly so; and when any of the elements are given the rest can easily be computed by the aid of this proportion, the precaution being of course taken to keep within the limits which have approved themselves most eligible in practice.

98. *Q.*—You stated that the capacity of the feed pump was a 240th of the capacity of the cylinder in the case of condensing engines—the engine being double acting and the pump single acting—and that in high pressure engines the capacity of the pump should be greater in proportion to the pressure of the

steam. Can you give any rule that will express the proper capacity for the feed pump at all pressures?

A.—That will not be difficult. In low pressure engines the pressure in the boiler may be taken at 5 lbs. above the atmospheric pressure, or 20 lbs. altogether; and as high pressure steam is merely low pressure steam compressed into a smaller compass, the size of the feed pump in relation to the size of the cylinder must obviously vary in the direct proportion of the pressure; and if it be a 240th of the capacity of the cylinder when the total pressure of the steam is 20 lbs., it must be a 120th of the capacity of the cylinder when the pressure is 40 lbs. per square inch, or 25 lbs. per square inch above the atmospheric pressure. This law of variation is expressed by the following rule:—multiply the capacity of the cylinder in cubic inches by the total pressure of the steam in lbs. per square inch, or the pressure per square inch on the safety valve plus 15, and divide the product by 4800; the quotient is the capacity of the feed pump in cubic inches, when the feed pump is single acting and the engine double acting. If the feed pump be double acting, or the engine single acting, the capacity of the pump must just be one-half of what is given by this rule.

99. *Q.*—How do you ascertain the power of high pressure engines?

A.—The actual power is readily ascertained by the indicator, by the same process by which the actual power of low pressure engines is ascertained. The friction of a locomotive engine when unloaded is found by experiment to be about 1 lb. per square inch on the surface of the pistons, and the additional friction caused by any additional resistance is estimated at about ·14 of that resistance; but it will be a sufficiently near approximation to the power consumed by friction in high pressure engines, if we make a deduction of a pound and a half from the pressure on that account, as in the case of low pressure engines. High pressure engines, it is true, have no air pump to work; but the deduction of a pound and a half of pressure is relatively a much smaller one where the pressure is high than where it does not much exceed the pressure of the atmosphere. The rule, therefore, for the actual horse power of a high pressure engine will stand thus:—square the diameter of the cylinder in inches, multiply by the pressure of the steam in the cylinder per square inch less 1½ lbs., and by the speed of the piston in feet per minute, and divide by 42,017; the quotient is the actual horse power.

100. *Q.*—But how do you ascertain the nominal horse power?

A.—The nominal horse power of a high pressure engine has never been defined; but it should obviously hold the same relation to the actual power as that which obtains in the case of condensing engines, so that an engine of a given nominal power may be capable of performing the same work, whether high pressure or condensing. This relation is maintained in the following rule, which expresses the nominal horse power of high pressure engines:—multiply the square of the diameter of the cylinder in inches by the pressure on the piston in pounds per square inch, and by the speed of the piston in feet per minute, and divide the product by 120,000; the quotient is the power of the engine in nominal horses power. If the pressure upon the piston be 80 lbs. per square inch, the operation may be abbreviated by multiplying the square of the diameter of the cylinder by the speed of the piston, and dividing by 1500, which will give the same result.

101. *Q.*—This rule for nominal horse power, however, is not representative of the dimensions of the cylinder. Cannot you give a rule for the nominal power of high pressure engines which shall discard altogether the element of velocity, and, as in the rule you have already given for the nominal power of low pressure engines, express merely the dimensions of the engine?

A.—Nothing is easier, if you maintain an unvarying pressure of steam; but as different pressures are used in different engines, the pressure must become an element in the computation. The rule for the nominal power will therefore stand thus:—multiply the square of the diameter of the cylinder in inches by the pressure on the piston in pounds per square inch, and by the cube root of the stroke in feet, and divide the product by 940; the quotient is the power of the engine in nominal horses power, the engine working at the ordinary speed of 128 times the cube root of the stroke.

102. *Q.*—Is 128 times the cube root of the stroke in feet per minute the ordinary speed of all engines?

A.—Locomotive engines travel at a quicker speed, —an innovation brought about, not by any process of scientific deduction, but by the accidents and exigencies of railway transit. All other engines, however, travel at about the speed of 128 times the cube root of the stroke in feet; and condensing engines, as at present constructed, cannot travel much quicker, as the valves of the air pump strike very hard and wear themselves quickly out if the engine is driven at a great velocity. To mitigate the shock in cases in which a higher speed has been desirable, as in the case of marine engines employed to drive the screw propeller without intermediate gearing, canvas air pump valves, resting on a perforated metal plate, have sometimes been adopted —the plies of canvas being either stuck together with India rubber or riveted with copper rivets, and being multiplied so as to make the valve half an inch in thickness; but valves of this kind, though they diminish the noise and tremor, wear rapidly away, and require to be renewed very often. It is very desirable, however, that this impediment should be completely surmounted, not merely as a means of meeting the new exigencies created by the application of the screw propeller to steam vessels, but because all engines will then be capable of working at a greatly increased speed, and with a proportionate increase of power. The most feasible way of enabling condensing engines to work at a high speed, appears to lie in the application of a slide valve to the air pump, instead of valves actuated by the pressure within the pump, and by performing the condensation in the air pump, instead of in the condenser, whereby the air pump valves will operate with as little noise at any speed as the cylinder valves, and the condenser may be dispensed with altogether. Engines constructed upon this plan may be driven at four times the speed of common engines, whereby an engine of large power may be purchased for a very moderate price, and be capable of being put into a very small compass; while the motion, from being more equable, will be better adapted for most purposes for which a rotary motion is required. Even for pumping mines and blowing iron furnaces, engines of this kind appear likely to come into use; for they are more suitable than other engines for driving the centrifugal pump, which in many cases appears likely to supersede other kinds of pumps for lifting water; and they are also conveniently applicable to the driving of fans, which, when so arranged that the air condensed by one fan is employed to feed another, and so on through a series of 4 or 5, have succeeded in forcing air into a furnace with a pressure of $2\frac{1}{2}$ lbs. on the square inch, and with a far steadier flow than can be obtained by a blast engine with any conceivable kind of compensating apparatus.

103. *Q.*—Then, if by this modification of the air pump you enable an engine to work at four times the speed, you also enable it to exert four times the power?

A.—Yes; always supposing it to be fully supplied with steam. The nominal power of this new species of engine, if condensing and if working at four times the ordinary speed, may be ascertained by the following rule:—multiply the square of the diameter of the cylinder in inches by the cube root of the stroke in feet, and divide the product by 12; the quotient is the power of the high speed condensing engine in nominal horses power. To ascertain the nominal power of a high pressure high speed engine, multiply the square of the diameter of the cylinder by the pressure on the piston per square inch less a pound and a half, and by the cube root of the stroke in feet, and divide the product by 235; the quotient is the power of the high pressure high speed engine in nominal horses power.

104. *Q.*—The high speed engine does not require so heavy a fly wheel as common engines?

A.—No: the fly wheel will be lighter, both by virtue of its greater velocity of rotation, and because the impulse communicated by the piston is less in amount and more frequently repeated, so as to approach more nearly to the condition of a uniform pressure. To find the proper quantity of cast iron for the rim of a fly wheel:—multiply the mean diameter of the rim by the number of its revolutions per minute, and square the product for a divisor; divide the number of nominal horses power of the engine by the number of strokes the piston makes per minute, multiply the quotient by the constant number 5,000,000, and divide the product by the divisor found as above; the quotient is the requisite quantity of cast iron in cubic feet to form the fly wheel rim.

105. *Q.*—What is the maximum speed with which a cast iron fly wheel may be driven without being burst asunder by its centrifugal force?

A.—The velocity of the rim should not exceed 60 feet per second, and at that speed the bursting force will amount to 1,100 lbs. on the square inch of section, setting aside the support derived from the arms. In engines for rolling iron, the fly wheel is driven usually at a very great speed, and it requires to be put together very firmly to resist the bursting force. The best practice in such cases is to cast the rim in a single piece, and to introduce malleable iron arms.

106. *Q.*—The steam is admitted to and from the cylinder by means of a slide or sluice valve?

A.—Yes; and of the slide valve there are many varieties; but the kinds most in use are the D valve, so called from its resemblance to a half cylinder or D in its cross section, and the three ported valve, which consists of a brass box set over the two ports or openings into the cylinder and a central port which conducts away the steam to the atmosphere or condenser; but the length of the box is so adjusted, that it can only cover one of the cylinder ports and the central or eduction port, at the same time. The effect, therefore, of moving the valve up and down, as is done by the eccentric, is to establish a connection alternately between each cylinder port and the passage whereby the steam escapes; and while the steam is escaping from beneath the piston, the position of the valve is such, that a free communication exists between the space above the piston and the steam in the boiler. The piston is thus urged alternately up and down, the valve so changing its position before the piston arrives at the end of the stroke, that the pressure is by that time thrown on the reverse side of the piston, so as to urge it into motion in the opposite direction. The valve does not move down when the piston moves down, nor does it move down when the piston moves up; but it moves from its mid position to the extremity of its throw, and back again to its mid position, while the piston makes an upward or downward movement, so that the motion is as it were at right angles to the motion of the piston; or it is the same motion that the piston of another engine, the crank of which is set at right angles with that of the first engine, would acquire.

107. *Q.*—What is meant by the lead of the valve?

A.—The amount of opening the valve presents for the admission of the steam, when the piston is just beginning its stroke. It is found expedient that the valve should have opened a little to admit steam on the reverse side of the piston before the stroke terminates; and the amount of this opening, which is given by turning the eccentric more or less round upon the shaft, is what is termed the lead.

108. *Q.*—And what is meant by the lap of the valve?

A.—It is an elongation of the valve face to a certain extent ever the port or towards the port, whereby the port is closed sooner than would otherwise be the case. This extension is chiefly effected at that part of the valve where the steam is admitted, or upon the *steam side* of the valve, as the technical phrase is; and the intent of the extension is to close the steam passage before the end of the stroke, whereby the engine is made to operate to a certain extent expansively. In some cases, however, there is also a certain amount of lap given to the eduction side, to prevent the eduction from being performed too soon when the lead is great; but in all cases there is far less lap on the eduction than on the steam side: very often there is none, and sometimes less than none, so that the valve is incapable of covering both the ports at once. The common stroke of the valve in rotative engines is twice the breadth of the port, and the length of the valve face will then be just the breadth of the port when there is lap on neither the steam nor eduction side. Whatever lap is therefore given, makes the valve face just so much longer. In some engines, however, the stroke of the valve is a good deal more than twice the breadth of the port; and it is by the stroke of the valve that the amount of lap is properly measurable.

109. *Q.*—Can you tell what amount of lap will accomplish any given amount of expansion?

A.—Yes, when the stroke of the valve is known. From the length of the stroke of the piston, subtract that part of the stroke which is intended to be accomplished before the steam is cut off; divide the remainder by the length of the stroke of the piston, and extract the square root of the quotient, which multiply by half the stroke of the valve, and from the product take half the lead; the remainder will be the lap required. To find how much before the end of the stroke the eduction passage will be closed:—to the lap on the steam side add the lead, and divide the sum by half the stroke of the valve; find the arc whose sine is equal to the quotient, and add 90° to it; divide the lap on the eduction side by half the stroke of the valve, and find the arc whose cosine is equal to the quotient; subtract this arc from the one last obtained, and find the cosine of the remainder; subtract this cosine from two, and multiply the remainder by half the stroke of the piston; the product is the distance of the piston from the end of the stroke when the eduction passage is closed. To find how far the piston is from the end of its stroke when the steam that is propelling it by expansion is allowed to escape to the atmosphere or condenser:—to the lap on the steam side add the lead, divide the sum by half the stroke of the valve, and find the arc whose sine is equal to the quotient; find the arc whose cosine is equal to the lap on the eduction side divided by half the stroke of the valve: add these two arcs together and subtract 90°; find the cosine of the residue, subtract it from one, and multiply the remainder by half the stroke of the piston; the product is the distance of the piston from the end of its stroke when the steam that is propelling it is allowed to escape into the atmosphere or condenser. In using these rules, all the dimensions are to be taken in inches, and the answers will be found in inches also.

110. *Q.*—To what extent can expansion be carried beneficially by means of lap upon the valve?

A.—To about one-third of the stroke; that is, the valve may be made with so much lap, that the steam will be cut off when two-thirds of the stroke have been performed, leaving the residue to be accomplished by the agency of the expanding steam; but if more lap be put on than answers to this amount of expansion, a very distorted action of the valve will be produced, which will impair the efficiency of the engine. If a further amount of expansion than this is wanted, it may be accomplished by wire-drawing the steam, or by so contracting the steam passage, that the pressure within the cylinder must decline when the speed of the piston is accelerated, as it is about the middle of the stroke. Thus, for example, if the valve be so made as to shut off the steam by the time two-thirds of the stroke have been performed, and the steam be at the same time throttled in the steam pipe, the full pressure of the steam within the cylinder cannot be maintained except near the beginning of the stroke where the piston travels slowly; for, as the speed of the piston increases, the pressure necessarily subsides, until the piston approaches the other end of the cylinder, where the pressure would rise again but that the operation of the lap on the valve by this time has had the effect of closing the communication between the cylinder and steam pipe, so as to prevent more steam from entering. By throttling the steam, therefore, in the manner here indicated, the amount of expansion due to the lap may be doubled, so that an engine with lap enough upon the valve to cut off the steam at two-thirds of the stroke, may, by the aid of wire-drawing, be virtually rendered capable of cutting off the steam at one-third of the stroke. The usual manner of cutting off the steam, however, is by means of a separate valve, termed an expansion valve; but such a device appears to be hardly necessary in ordinary engines. In the Cornish engines, where the steam is cut off in some cases at one-twelfth of the stroke, a separate valve for the admission of steam, other than that which permits its escape, is of course indispensable; but in common rotative engines, which may realize expansive efficacy by throttling, a separate expansive valve does not appear to be required.

111. *Q.*—What size of orifice is allowed for the escape of the steam through the safety valve?

A.—About 0·8 of a circular inch per horse power in condensing engines, or a circular inch per 1¼ horse power. The following rule, however, will give the dimensions suitable for all kinds of engines, whether high or low pressure:—multiply the square of the diameter of the cylinder in inches by the speed of the piston in feet per minute, and divide the product by 375 times the pressure on the boiler per square inch;

the quotient is the proper area of the safety valve in square inches. This rule of course supposes that the evaporating surface has been properly proportioned to the engine power?

112. *Q.*—Will you state the proper dimensions of the air pump and condenser and of the steam passages?

A.—Mr. Watt made the air pump of his engine half the diameter of the cylinder and half the stroke, or one-eighth of the capacity, and the condenser was usually made about the same size as the air pump; but as the pressure of the steam has been increased in all modern engines, it is better to make the air pump a little larger than this proportion. 0·6 of the diameter of the cylinder and half the stroke answers very well, and the condenser may be made as large as it can be got with convenience, though the same size as the air pump will suffice. The common size of the cylinder passages is one-twenty-fifth of the area of the cylinder, or one-fifth of the diameter of the cylinder, which is the same thing. This proportion corresponds very nearly with one square inch per horse power when the length of the cylinder is about equal to its diameter; and one square inch of area per horse power for the cylinder ports and eduction passages answers very well in the case of engines working at the ordinary speed of 220 feet per minute. The area of the steam pipe is usually made less than the area of the eduction pipe, especially when the engine is worked expansively, and with a considerable pressure of steam. In the case of ordinary condensing engines, however, working with the usual pressure of from 4 to 8 lbs. above the atmosphere, the area of the steam pipe is not less than a circular inch per horse power; in such engines the diameter of the steam pipe may be found by the following rule:—divide the number of nominal horse power by 0·8, and extract the square root of the quotient, which will be the internal diameter of the steam pipe. The area of the injection orifice should be about one-fifteenth of a square inch per nominal horse power; but the area of the injection pipe should be a little larger: a tenth of a circular inch per horse power answers very well, or the diameter of the injection pipe may be found by dividing the nominal horse power by 10, and extracting the square root of the quotient, which will be the diameter in inches. The area through the foot and discharge valves is usually made equal to one-fourth of the area of the air pump, and the diameter of the waste water pipe is made one-fourth of the diameter of the cylinder, which gives an area somewhat less than that of the foot and discharge valve passages.

113. *Q.*—Will you explain by what process of computation these proportions are arrived at?

A.—The size of the steam pipe is so regulated that there will be no material disparity of pressure between the cylinder and boiler, and in fixing the size of the eduction passage the same object is kept in view. When the diameter of the cylinder and the velocity with which the piston travels are known, it is easy to tell what the velocity of the steam in the steam pipe will be; for if the area of the cylinder be 25 times greater than that of the steam pipe, the steam in the steam pipe must travel 25 times faster than the piston, and the difference of pressure requisite to produce this velocity of the steam can easily be ascertained, by finding what height a column of steam must be to give that velocity, and what the weight or pressure is of such a column. The proper area of the injection orifice per horse power can easily be told when the quantity of water requisite to condense the steam is known, and the pressure specified under which the water enters the condenser. The vacuum in the condenser may be taken at 26 in. of mercury, which is equivalent to a column of water 29·4 ft. high, and the square root of 29·4 multiplied by 8·021 is 43·15, which is the velocity in feet per second that a heavy body would acquire in falling 29·4 ft., or with which the water would enter the condenser. Now, if a cubic foot of water evaporated per hour be equivalent to an actual horse power, and 28·9 cubic inches of water be requisite for the condensation of a cubic inch of water in the form of steam, 28·9 cubic feet of condensing water per actual horse power per hour for the engine, or 13·905 cubic inches per second will be necessary, and the size of the injection orifice must be such that this quantity of water flowing with the velocity of 43·15 ft. per second, or 517·8 inches per second, will gain admission to the condenser. Dividing, therefore, 13·905, the number of cubic inches to be injected, by 517·8, the velocity of influx in inches per second, we get 0·02685 for the area of the orifice in square inches; but inasmuch as it has been found by experiment that the actual discharge of water through a hole in a thin plate is only six-tenths of the theoretical discharge on account of the contracted vein, the area of the orifice must be increased in the proportion of such diminution of effect, or be made 0·04475, or $\frac{1}{22}$d of a square inch per horse power. This, it will be remarked, is the area required per actual horse power; in most of the modern engines there is more than a cubic foot of water evaporated per hour per nominal horse power, and it is therefore found expedient to make the area of the injection orifice a third larger than the area due to the actual power, or $\frac{1}{15}$th of a square inch per nominal horse power, as already prescribed.

114. *Q.*—If the relation you have mentioned subsist between the area of the steam passages and the velocity of the piston, then the passages must be larger when the piston travels very rapidly?

A.—And they are so made. The area of the ports of locomotive engines is usually so proportioned as to be from $\frac{1}{10}$th to $\frac{1}{8}$th the area of the cylinder—in some cases even as much as $\frac{1}{6}$th; and in all high speed engines the ports should be very large, and the valve should have a good deal of travel. The area of port which it appears advisable to give to modern engines of every description, is expressed by the following rule:—multiply the area of the cylinder in square inches by the speed of the piston in feet per minute, and divide the product by 4,000; the quotient is the area of each cylinder port in square inches. This rule gives rather more than a square inch of port per nominal horse power to condensing engines working at the ordinary speed; but the excess is but small, and is upon the right side. In locomotive engines the eduction pipe passes into the chimney, and the force of the issuing steam has the effect of maintaining a rapid draught through the furnace. The orifice of the waste steam pipe, or the blast pipe as it is termed, is much contracted in some engines with a view of producing a fiercer draught; but an area of $\frac{1}{22}$d of the area of the cylinder is a common proportion.

115. Q.—If the area of the blast pipe be much diminished, the escaping steam must so far resist the motion of the piston as to diminish considerably the power of the engine?

A.—At a speed of 10 miles an hour the resistance to the piston occasioned by the blast pipe may be taken on the average at about 2 lbs. per square inch, and the resistance increases directly as the speed of the engine. In some cases, however, the blast pipe is so much contracted, that as much as half the power of the engine is spent in forcing the steam through it. The object of this undue contraction is to increase the intensity of the draught in the furnace, and as a consequence to increase the production of steam; but engines in which such a waste of power exists from this single cause, are altogether unsuitable for the work they have to do, in consequence of their insufficient evaporating power. The question of draught is only a question of area of fire grate, and it is much preferable to enlarge the fire grate than to increase the intensity of the draught. The area of the chimney in locomotives is usually made $\frac{1}{10}$th of the area of the fire grate, and its height must not exceed 14 ft. above the level of the rails, else it will come in contact with the bridges which cross the line. A given area of heating surface in locomotive boilers is more effective than the same area of heating surface in other boilers, on account of the greater heat to which it is subjected. If, at a speed of 20 miles an hour, a locomotive boils off a cubic foot of water in the hour for every six square feet of heating surface, as appears to be an average performance, then 1 square foot of heating surface must boil off 10·4 lbs. of water per hour, at a speed of 20 miles an hour, and the rate of evaporation will vary nearly as the fourth root of the speed.

116. *Q.*—What is the amount of tractive force requisite to draw carriages on railways?

A.—Upon well formed railways, with carriages of good construction, the average tractive force required for low speeds is about 7½ lbs. per ton, or $\frac{1}{300}$th of the load; though in some experimental cases, where particular care was taken to obtain a favorable result, the tractive force has been reduced as low as $\frac{1}{500}$th of the load. At low speeds the whole of the tractive force is expended in overcoming the friction, which is made up partly of the friction of attrition in the axles, and partly of the rolling friction, or the obstruction to the rolling of the wheels upon the rail. The rolling friction is very small when the surfaces are smooth, and in the case of railway carriages does not exceed $\frac{1}{1000}$th of the load; whereas the draught on common roads of good construction, which is chiefly made up of the rolling friction, is as much as $\frac{1}{36}$th of the load.

117. *Q.*—In reference to friction you have already stated, in your answer to Question 33, that the friction of iron sliding upon brass, which has been oiled and then wiped dry, so that no film of oil is interposed, is about $\frac{1}{11}$th of the pressure, but that in machines in actual operation, where there is a film of oil between the rubbing surfaces, the friction is only about one-third of this amount, or $\frac{1}{33}$d of the weight. How then can

the tractive resistance of locomotives at low speeds, which you say is entirely made up of friction, be so little as $\frac{1}{500}$th of the weight?

A.—I did not state that the resistance to traction was $\frac{1}{500}$th of the weight upon an average—to which condition the answer given to Question 33 must be understood to apply—but I stated that the average traction was about $\frac{1}{300}$th of the load, which nearly agrees with my former statement. If the total friction be $\frac{1}{300}$th of the load, and the rolling friction be $\frac{1}{1000}$th of the load, then the friction of attrition must be $\frac{1}{429}$th of the load; and if the diameter of the wheels be 36 in., and the diameter of the axles be 3 in., which are common proportions, the friction of attrition must be increased in the proportion of 36 to 3, or 12 times, to represent the friction of the rubbing surface when moving with the velocity of the carriage. $\frac{12}{429}$ths are about $\frac{1}{35}$th of the load, which does not differ much from the proportion of $\frac{1}{33}$d, as previously determined. While this, however, is the average result, the friction is a good deal less in some cases. Mr. Southern, in some experiments upon the friction of the axle of a grindstone—an account of which may be found in the 65th volume of the "Philosophical Transactions"—found the friction to amount to less than $\frac{1}{40}$th of the weight; and Mr. Wood, in some experiments upon the friction of locomotive axles, found that by ample lubrication the friction might be made as little as $\frac{1}{60}$th of the weight, and the traction, with the ordinary size of wheels, would, in such a case, be about $\frac{1}{500}$th of the weight. The function of lubricating substances is to prevent the rubbing surfaces from coming into contact, whereby abrasion would be produced, and unguents are effectual in this respect in the proportion of their viscidity; but if the viscidity of the unguent be greater than what suffices to keep the surfaces asunder, an additional resistance will be occasioned; and the nature of the unguent selected should always have reference, therefore, to the size of the rubbing surfaces, or to the pressure per square inch upon them. With oil the friction appears to be a minimum when the pressure on the surface of a bearing is about 90 lbs. per square inch: the friction from too small a surface increases twice as rapidly as the friction from too large a surface, added to which, the bearing, when the surface is too small, wears rapidly away.

118. *Q.*—What is the amount of adhesion of the wheels upon the rails?

A.—The adhesion of the wheels upon the rails is about $\frac{1}{5}$th of the weight when the rails are clean, or either perfectly wet or perfectly dry; but when the rails are half wet or greasy, the adhesion is not more than $\frac{1}{10}$ or $\frac{1}{12}$th of the weight or pressure upon the wheels. The weight of a locomotive of modern construction varies from 20 to 25 tons.

119. *Q.*—And what is its cost and average performance?

A.—The cost of a common narrow gauge locomotive at the present time varies from 1900*l.* to 2200*l.*; it will run on an average 130 miles per day, at a cost for repairs of 2½*d.* per mile; and the cost of locomotive power, including repairs, wages, oil, and coke, does not much exceed 6*d.* per mile run, on economically managed railways. This does not include a sinking fund for the renewal of the engines when worn out, which may be taken as equivalent to 10 per cent. on their original cost.

120. *Q.*—Does the expense of traction increase much with an increased speed?

A.—Yes; it increases very rapidly, partly from the undulation of the earth when a heavy train passes over it at a high velocity, but chiefly from the resistance of the atmosphere, which constitutes the greatest of the impediments to motion at high speeds. At a speed of 30 miles an hour, the atmospheric resistance amounts to about 12 lbs. a ton, and in side winds the resistance even exceeds this amount, partly in consequence of the additional friction caused from the flanges of the wheels being forced against the rails, and partly because the wind catches to a certain extent the front of every carriage, whereby the frontage exposed to the wind is virtually increased. At a speed of 30 miles an hour, an engine evaporating 200 cubic feet of water in the hour, and therefore exerting about 200 horses power, will draw a load of 110 tons. Taking the friction of the train at 7½ lbs. per ton, or 825 lbs. operating at the circumference of the driving wheel,—which, with 5 ft. 6 in. wheels and 18 in. stroke, is equivalent to 4757 lbs. upon the piston,—and taking the resistance of the blast pipe at 6 lbs. per square inch of the pistons, and the friction of the engine unloaded at 1 lb. per square inch, which, with pistons 12 in. in diameter, amount together to 1582 lbs., and reckoning the increased friction of the engine due to the load at $\frac{1}{7}$th of the load, as in some cases it has been found experimentally to be, though a much less proportion than this would probably be a nearer average, we have 7018·4 lbs. for the total load upon the pistons. At 30 miles an hour the speed of the pistons will be at 457·8 feet per minute, and 7018·4 lbs. multiplied by 457·8 ft. per minute, are equal to 3213023·5 lbs. raised 1 foot high in the minute, which, divided by 33,000, gives 97·3 horses power as the power which would draw 110 tons upon a railway at a speed of 30 miles an hour, if there were no atmospheric resistance. The atmospheric resistance, with a load of 110 tons, is at the rate of 12 lbs. a ton, equal to 1320 lbs., moving at a speed of 30 miles an hour, which, when reduced, becomes 105·8 horses power, and this, added to 97·3, makes 203·1, instead of 200 horses power, as ascertained by a reference to the evaporative power of the boiler, which small excess of efficiency is probably produced by a certain amount of expansive action in the engine. The atmospheric resistance is found to increase as the square of the velocity, while the other resistances do not increase with the velocity at all per unit of space passed over, but increase as the velocity per unit of time during which they act. It is not difficult, therefore, to approximate to the power requisite for the propulsion of a train at any rate of speed, when the distribution of power at any one speed has been ascertained. At a speed of 60 miles an hour, for example, the power required to overcome the friction of a train weighing 110 tons, will be $97·3 \times 2 = 194·6$ horses power; while the power requisite to overcome the atmospheric resistance will be $105·8 \times 4 = 423·2$ horses power, if the same distance be passed over at the speed of 30 miles an hour; but as twice the distance will be passed over in the hour when the speed is twice as great, the power requisite to overcome the atmospheric resistance will be twice this amount, or 846·4 horses power, making together a power of 1041 horses to draw a train weighing 110 tons, at a speed of 60 miles an hour. A locomotive, to perform this duty, should have 6246 square feet of heating surface in the boiler, and the cylinders should be sufficiently capacious to permit 320,628 cubic feet of steam to pass through them in the hour, though it will be preferable to make the capacity considerably greater than this, and to work the steam expansively.

121. *Q.*—How comes it, that the resistance of the air to the motion of a locomotive increases as the square of the velocity, instead of the velocity simply?

A.—Because the height necessary to generate the velocity with which the train strikes the air, or the air strikes the carriage, increases as the *square* of the velocity, and the resistance or the weight of a column of air, or of any other fluid, varies as the height. A falling body, as has been already explained in the answer to Question 14, to have acquired twice the velocity, must have fallen through four times the height; the velocity generated by a column of any fluid is equal to that acquired by a body falling through the height of the column; and it is therefore clear, that the pressure due to any given velocity, must be as the square of that velocity, the pressure being in every case directly as the altitude of the column. The work done, however, by a stream of air or other fluid in a given time, will vary as the cube of the velocity; for if the velocity of a stream of air be doubled, there will not only be four times the pressure exerted per square foot, but twice the quantity of air will be employed; and in windmills, accordingly, it is found, that the work done varies nearly as the cube of the velocity of the wind. If, however, the work done by *a given quantity* of air moving at different speeds be considered, it will vary as the squares of the speeds.

122. *Q.*—But in a locomotive there is no work done by the wind, and as the resistance of the air upon the train varies as the square of the speed, should not the power requisite to overcome that resistance vary as the square of the speed?

A.—It should, if you consider the resistance over a given distance, and not the resistance during a given time. It will take four times the power, so far as atmospheric resistance is concerned, to accomplish a mile at the rate of 60 miles an hour that it will take to accomplish a mile at 30 miles an hour; but in the former case there will be twice the number of miles accomplished in the same time, so that when the velocity of a train is doubled, we require an engine that is capable of overcoming four times the resistance at twice the speed; or, in other words, that is capable of exerting eight times the power, so far as regards the element of atmospheric resistance.

123. *Q.*—When a train moves at the rate of 60 miles an hour, what will be the centrifugal force of the wheels?

A.—60 miles an hour is 88 ft. per second, and if the centrifugal force of a wheel travelling at this speed be computed by the rule given in Answer 24, it will be found to amount to 80·4 times the weight of the wheel, or 4124·5 lbs. per square inch of sectional area of the rim of the wheel, supposing any one point of the rim to sustain all the centrifugal force produced by its own revolution. About 4000 lbs. per square inch of sectional area is the utmost strain to which iron should be exposed in machinery, and even if we take the centrifugal force as divided between the two sides of the rim, railway wheels can scarcely be considered safe at a speed of 60 miles an hour, unless so constructed that the centrifugal force of the rim will be counteracted, to a material extent, by the centripetal action of the arms. Hooped wheels are very unsafe, unless the hoops are, by some process or other, firmly attached to the arms. It is of no use to increase the dimensions of the rim of a wheel with a view of giving increased strength to counteract the centrifugal force, as every increase in the dimensions of the rim will increase the centrifugal force in the same proportion.

124. *Q.*—Does the law of resistance to the motion of railway trains extend to the case of vessels moving in water?

A.—The resistance of vessels moving in water varies as the square of the velocity, and the power required to overcome that resistance as the cube of the velocity; so that the resistance to ships moving in water varies in the same ratio as the resistance presented by the atmosphere to the carriages in the case of railway trains. To double the velocity of a steamer it is necessary that the engines should exert eight times the power; but a double velocity may be gained without any increase of power by such a modification of the form of the vessel as will enable it to pass through the water with greater facility. It is impossible, therefore, in the case of a steam vessel, to tell what velocity with a given load a given power of engine will produce, unless the shape of the vessel be also known ; and the shapes of vessels are so various and inconsistent, that it is impossible to find any one expression by which they may all be represented.

125. *Q.*—But may not the different shapes be so classified that the speed answerable to a given power in a specified class can be approximately predicted ?

A.—Yes, that may be done, and *has* been done, by Boulton and Watt, who have ascertained, experimentally, the speed realized by different forms of vessels with different powers of engine, and have deduced from thence rules which enable them to tell with great precision the speed which any new vessel, of a particular form and power, will achieve. The first set of experiments was made in 1828, upon the vessels Caledonia, Diana, Eclipse, Kingshead, Moordyke, and Eagle—vessels of a similar form and all with square bilges and flat floors ; and the result was to establish the number 925 as the co-efficient of performance of such vessels. This co-efficient is obtained by multiplying the cube of the velocity of the vessel in miles per hour by the sectional area of immersed midship-section in square feet, and dividing by the nominal horse power ; and its use is to enable us to determine the speed of any similar vessel with any other area of midship-section, and any other number of nominal horses power. The better the shape of the vessel is, the larger the co-efficient becomes, as appears by the second set of experiments, which were made upon the superior vessels Venus, Swiftsure, Dasher, Arrow,

Spitfire, Fury, Albion, Queen, Dart, Hawk, Margaret, and Hero—all vessels having flat floors and round bilges, where the co-efficient became 1160. The third set of experiments was made upon the vessels Lightning, Meteor, James Watt, Cinderella, Navy Meteor, Crocodile, Watersprite, Thetis, Dolphin, Wizard, Escape, and Dragon—all vessels with rising floors and round bilges, and the co-efficient of performance was found to be 1430. The fourth set of experiments was made in 1834, upon the vessels Magnet, Dart, Eclipse, Flamer, Firefly, Ferret, and Monarch, when the co-efficient of performance was found to be 1580. The velocity of any of these vessels, with any power of sectional area, may be ascertained by multiplying the co-efficient of its class by the nominal horse power, dividing by the sectional area in square feet, and extracting the cube root of the quotient, which will be the velocity in miles per hour; or the number of nominal horses power requisite for the accomplishment of any required speed may be ascertained by multiplying the cube of the required velocity in miles per hour, by the sectional area in square feet, and dividing by the co-efficient: the quotient is the number of nominal horse power requisite to realize the speed. In the whole of these experiments the pressure of steam in the boiler varied between 2¾ lbs. and 4 lbs. per square inch, and the effective pressure on the piston varied between 11 lbs. and 13 lbs. per square inch, so that the average ratio of the nominal to the actual power may be easily computed; but it will be preferable to state the nominal power of some of the vessels, and their actual power as ascertained by experiment. Of the

Eclipse, the nominal power was 76, and the actual power 144·4, horses; of the Arrow, the nominal power was 60, and the actual 119·5; Spitfire, nominal 40, actual 64; Fury, nominal 40, actual 65·6; Albion, nominal 80, actual 135·4; Dart, nominal 100, actual 152·4; Hawk, nominal 40, actual 73; Hero, nominal 100, actual 171·4; Meteor, nominal 100, actual 160; James Watt, nominal 120, actual 204; Watersprite, nominal 76, actual 157·6; Dolphin, nominal 140, actual 238; Dragon, nominal 80, actual 131; Magnet, nominal 140, actual 238; Dart, nominal 120, actual 237; Flamer, nominal 120, actual 234; Firefly, nominal 52, actual 86·6; Ferret, nominal 52, actual 88; Monarch, nominal 200, actual 378. In the case of swift vessels of modern construction, such as the Red Rover, Herne, Queen, and Prince of Wales, the co-efficient appears to be about 2550; but in these vessels there is a still greater excess of the actual over the nominal power than in the case of the vessels previously enumerated, and the increase in the co-efficient is consequent upon the increased pressure of the steam in the boiler, as well as the superior form of the ship. The nominal power of the Red Rover, Herne, and City of Canterbury is, in each case, 60 horses; but the actual power of the Red Rover is 147, of the Herne 177, and of the City of Canterbury 153, and in some vessels the excess is still greater; so that with such variations it becomes necessary to adopt a co-efficient derived from the introduction of the actual instead of the nominal power. In the first class of vessels experimented upon, the actual power was about 1·6 times greater than the nominal power; in the second class, 1·67 times greater; in the third class, 1·7 times greater; and in the fourth, 1·96 times greater; while in such vessels as the Red Rover and City of Canterbury, it is 2·65 times greater; so that if we adopt the actual instead of the nominal power in fixing the co-efficients, we shall have 554 as the first co-efficient, 694 as the second, 832 for the third, and 806 for the fourth, instead of 925, 1160, 1430, and 1580, as previously specified; while for such vessels as the Red Rover, Herne, Queen, and Prince of Wales, we shall have 962 instead of 2550. These smaller co-efficients then express the relative merits of the different vessels without reference to any difference of efficacy in the engines, and it appears preferable, with such a variable excess of the actual over the nominal power, to employ them instead of those first referred to. From the circumstance of the third of the new co-efficients being greater than the fourth, it appears that the superior result in the fourth set of experiments arose altogether from a greater excess of the actual over the nominal power.

126. *Q.*—In what way is it that the shape of a vessel influences her speed, since vessels of the same sectional area must manifestly put in motion a column of water of the same magnitude, and with the same velocity?

A.—A vessel will not strike the water with the same velocity when the bow lines are sharp as when they are otherwise, for a very sharp bow has the effect of enabling the vessel to move through a great distance while the particles of water are moved aside but a small distance; or, in other words, it causes the velocity with which the water is moved to be very small relatively with the velocity of the vessel; and as the resistance increases as the square of the velocity with which the water is moved, it is conceivable enough in what way a sharp bow may diminish the resistance. In vessels in which a high rate of speed is intended to be realized, it will generally be advantageous to put the midship frame abaft the centre of the vessel; or, in other words, to make the bow end sharper than the stern end, although the stern end must be sharp also, else the water will be unable to fill the vacuity behind the vessel with sufficient rapidity to obviate the resistance due to a difference of level, and the vessel also will steer badly. It appears expedient, in most cases, to make the shape of the bow lines such, that in equal times the particles of water shall occasion equal increments of resistance; and it is also most important to keep vessels intended to be swift as light as possible, as the difference even of a few tons in the weight may materially affect the speed.

127. *Q.*—Will a vessel experience more resistance in moving in salt water than in moving in fresh?

A.—If the immersion be the same in both cases a vessel will experience more resistance in moving in salt water than in moving in fresh, on account of the greater density of salt water; but as the flotation is proportionably greater in the salt water, the resistance will be the same with the same weight carried. The resistance opposed to any body moving in a fluid is directly porportional to the quantity of matter moved, and the height necessary to generate the velocity of motion; but the quantity of matter moved in any given distance is directly as the density or specific gravity of the fluid; so that the density multiplied by the height requisite to generate the velocity, or, what is the same thing, by the velocity squared, will express the resistance universally. It has already been stated, in the answer to Question 13, that the square root of the height from which a body falls in feet multip ied by 8·021, will give its velocity of motion in feet per second; therefore the height multiplied by the square of 8·021, or 64⅓, will give the square of the velocity in feet per second, and the square of the velocity divided by 64⅓ will give the height.

128. *Q.*—Then the resistance experienced by a vessel in passing through the water, will be less than that of a flat board of the same area as the cross section of the vessel?

A.—Yes, very much less, as is illustrated by the circumstance that steam vessels are propelled by a very small area of paddle float in proportion to their sectional area. When a flat board is moved through a fluid with its flat side foremost, the resistance it suffers is equal to the weight of a column of the fluid with a base of the same area as the board, and with the altitude due to the velocity of motion with which the board moves; and the resistance will be the same whether the board moves through a quiescent fluid or a moving fluid strikes against a stationary board. As the altitude in feet due to the velocity of motion in feet per second, is equal to the square of the velocity divided by 64⅓, and as a cubic foot of fresh water weighs 62½ lbs., the pressure per square foot in lbs. upon a board moved through river water will be 62½ times the square of the velocity in feet per second divided by $64\frac{1}{3}$; or it will be the square of the velocity multiplied by ·9715. To express the pressure or resistance in actual horses power, we must multiply this quantity by 60 times the velocity in feet per second, which will give the velocity in feet per minute, and divide by 33,000 for horses power, which makes the resistance per square foot in horses power equal to the cube of the velocity in feet per second multiplied by ·00176637. If the water strikes the board obliquely, the resistance will follow a different law. When a body impinges obliquely upon a plane, the force of impact with any given velocity varies as the sine of the angle of incidence; and therefore the force with which the particles of water strike against a board will vary as the sine of the angle at which they strike it—becoming of course less and less as the board is turned more upon its edge. But the number of particles striking the board will also vary as the sine of the angle of incidence, or, in other words, as the perpendicular height of the inclined board; so that the resistance, as it varies both with the force with which the particles strike the board, and the number of particles which strike it, must vary as the square of the sine of the angle of incidence. The horizontal resistance, therefore, per square foot in horses power of a board which is struck by the water obliquely, such as the float of a water wheel, or which strikes the water obliquely, such as the float of a paddle wheel, will be found by the following rule:—multiply the cube or third power of the velocity in feet per second by the area of the plane in feet, by the square of the sine of the angle of incidence, and by the constant co-efficient ·00176637; the result will be the resistance in actual horses power of a board striking the water obliquely or obliquely struck by the water.

129. *Q.*—But in a paddle wheel the floats do not move in a horizontal line?

A.—There are two kinds of paddle wheels in extensive use, the one being the ordinary radial wheel, in which the floats are fixed on arms radiating from the centre, and the other the feathering wheel, in which each float is hung upon a centre, and is so governed by suitable mechanism as to be always kept in nearly the vertical position. In the radial wheel there is some loss of power from oblique action, whereas in the feathering wheel there is little or no loss from this cause; but in every kind of paddle there is a loss of power from the recession of the water from the float boards, or the *slip* as it is commonly called; and this loss is the necessary condition of the resistance for the propulsion of the vessel being created in a fluid. The slip is expressed by the difference between the speed of the wheel and the speed of the vessel; and the larger this difference is, the greater the loss of power from slip must be—the consumption of steam in the engine being proportionate to the velocity of the wheel, and the useful effect being proportionate to the speed of the ship. In the feathering wheel, where every part of any one immerged float moves forward with the same horizontal velocity, the pressure or resistance may be supposed to be concentrated near the centre of the float; whereas in the common radial wheel this cannot be the case; for as the outer edge of the float moves more rapidly than the edge nearest the centre of the wheel, the outer part of the float is the most effectual in propulsion. The point at which the outer and inner portions of the float just balance one another in propelling effect, is called the *centre of pressure;* and if all the resistances were concentrated in this point, they would have the same effect as before in propelling the vessel. The resistance upon any one moving float board totally immersed in the water will, when the vessel is at rest, obviously vary as the square of its distance from the centre of motion—the resistance of a fluid varying with the square of the velocity; but, except when the wheel is sunk to the axle or altogether immersed in the water, it is impossible, under ordinary circumstances, for one float to be totally immersed without others being immersed partially, whereby the arc described by the extremity of the paddle arm will become greater than the arc described by the inner edge of the float; and consequently the resistance upon any part of the float will increase in a higher ratio than the square of its distance from the centre of motion—the position of the centre of pressure being at the same time correspondingly affected. In the feathering wheel the position of the centre of pressure of the entering and emerging floats is continually changing from the lower edge of the float—where it is when the float is entering or leaving the water—to the centre of the float, which is its position when the float is wholly immersed; but in the radial wheel, the centre of pressure can never rise so high as the centre of the float.

130. *Q.*—All this relates to the action of the paddle

when the vessel is at rest; will you explain its action when the vessel is in motion?

A.—When the wheel of a coach rolls along the ground, any point of its periphery describes in the air a curve which is termed a cycloid; any point within the periphery traces a prolate or protracted cycloid, and any point exterior to the periphery traces a curtate or contracted cycloid—the prolate cycloid partaking more of the nature of a straight line, and the curtate cycloid more of the nature of a circle. The action of a paddle wheel in the water resembles in this respect that of the wheel of a carriage running along the ground: that point in the radius of the paddle, of which the rotative speed is just equal to the velocity of the vessel, will describe a cycloid; points nearer the centre, prolate cycloids; and points further from the centre, curtate cycloids. The circle described by the point whose velocity equals the velocity of the ship, is called the *rolling circle;* and the resistance due to the difference of velocity of the rolling circle and centre of pressure, is that which operates in the propulsion of the vessel. The resistance upon any part of the float, therefore, will vary as the square of its distance from the rolling circle, supposing the float to be totally immersed; but, taking into account the greater length of time during which the extremity of the paddle acts, whereby the resistance will be made greater, we shall not err far in estimating the resistance upon any point at the third power of its distance from the rolling circle in the case of light immersions, and the 2·5 power in the case of deep immersions. With this assumption, which accords sufficiently with experiment to justify its acceptation, the position of the centre of pressure may be found by the following rule:—from the radius of the wheel subtract the radius of the rolling circle; to the remainder add the depth of the paddle board, and divide the fourth power of the sum by four times the depth; from the cube root of the quotient subtract the difference between the radii of the wheel and rolling circle, and the remainder will be the distance of the centre of pressure from the upper edge of the paddle. The diameter of the rolling circle is very easily found, for we have only to divide 5280 times the number of miles per hour, by 60 times the number of strokes per minute, to get an expression for the circumference of the rolling circle: or the following rule may be adopted:—divide 88 times the speed of the vessel in statute miles per hour, by 3·1416 times the number of strokes per minute; the quotient will be the diameter in feet of the rolling circle. The diameter of the circle in which the centre of pressure moves, or the effective diameter of the wheel being known, and also the diameter of the rolling circle, we at once find the excess of velocity of the wheel over the vessel, which, multiplied by $\frac{62\frac{1}{2}}{64\frac{1}{2}}$, or, ·9715, will give the resistance per square foot upon the vertical paddle.

131. *Q.*—Will you illustrate these rules by an example?

A.—A steam vessel of moderately good shape, and with engines of 200 horses power, realizes, with 22 strokes per minute, a speed of 10·62 miles per hour. To find the diameter of the rolling circle, we have 88 times 10·62, equal to 934·66, and 22 times 3·1416, equal to 69·1152; then 934·66 divided by 69·1152, is equal to 13·52 feet, which is the diameter of the rolling circle. The diameter of the wheel is 19 ft. 4 in., so that the diameter of the rolling circle is about ⅔ds of the diameter of the wheel, and this is a frequent proportion. The depth of the paddle board is 2 feet, and the difference between the diameters of the wheel and rolling circle will be 5·8133, which will make the difference of their radii 2·9067; and, adding to this the depth of the paddle board, we have 4·9067, the fourth power of which is 579·74, which, divided by four times the depth of the paddle board, gives us 72·455, the cube root of which is 4·1689, which, diminished by the difference of the radii of the wheel and rolling circle, leaves 1·2622 feet for the distance of the centre of pressure from the upper edge of the paddle board in the case of light immersions. The radius of the wheel being 9·6667, the distance from the centre of the wheel to the upper edge of the float is 7·6667, and adding to this 1·2622, we get 8·9299 feet as the radius, or 17·8598 feet as the diameter of the circle in which the centre of pressure revolves. With 22 strokes per minute, the velocity of the centre of pressure will be 20·573 feet per second, and with 10·62 miles per hour for the speed of the vessel, the velocity of the rolling circle will be 15·576 feet per second. The effective velocity will be the difference between these quantities, or 4·997 feet per second; and taking the length of the floats at 10 feet, which makes the area of each float 20 square feet, the resistance in lbs. encountered by the vertical float will be 20 multiplied by $4·997^2$, multiplied by 0·9715, equal to 485·17 lbs., which being doubled for the vertical float of the other wheel, gives us 970·34 lbs. as the pressure on the vertical floats in their motion through the water. The velocity with which the floats move is that of the centre of pressure, so that to find the resistance in horses power, we must multiply the velocity of the centre of pressure in feet per second by 60, to bring it to feet per minute, then by 970·34 lbs., and divide by 33,000, which will give about 36 horses power as the power expended on the vertical paddles. If the nominal power of the engines be 200 horses, the actual horses power will be greater in engines of this class in the proportion of 33,000 to 50,000; so that the actual power exerted by the engines will be about 315 horses, and 0·114 of the power of the engines will be the proportion of power expended upon the vertical paddles.

132. *Q.*—That is in the case of a common radial wheel: what will be the proportion in the feathering wheel?

A.—It will be the same in the feathering wheel, if the float be of the same area, and all other circumstances the same; but in the radial wheel the vertical float sustains the least resistance in propulsion, and in the feathering wheel the most resistance; and as the power of the engine is a fixed quantity, which has to be expended among all the floats, and as in the radial wheel it is chiefly distributed among the oblique floats which encounter a resistance that has to be thrown in a great measure upon the vertical float in the case of the feathering wheel, a larger area of float-board is necessary in the feathering wheel than in the radial one. To understand how the entering and emerging paddles of the radial wheel sustain more resistance in propulsion than the vertical one, it is necessary to remember that the only resistance upon the vertical paddle is that due to the difference of velocity of the wheel and the ship; but if the wheel be supposed to be immersed to its axle, so that the entering float strikes the water horizontally, it is clear that the resistance on such float is that due to the whole velocity of rotation; and that the resistance to the entering float will be the same whether the vessel is in motion or not. The resistance opposed to the rotation of any float increases from the position of the vertical float—where the resistance is that due to the difference of velocity of the wheel and vessel—until it reaches the plane of the axis, supposing the wheel to be immersed so far, where the resistance is that due to the whole velocity of rotation; and although in any oblique float the total resistance cannot be considered operative in a horizontal direction, yet the total resistance increases so rapidly on each side of the vertical float, that the portion of it which is operative in the horizontal direction, is in all ordinary cases of immersion very great when compared with the whole resistance upon the vertical float. In the feathering wheel, where there is none of this oblique action, the resistance will be simply in the proportion of the square of the horizontal velocities of the several floats, which may be represented by the horizontal distances between them; and in the feathering wheel, the vertical float having the greatest horizontal velocity will have the greatest propelling effect.

133. *Q.*—Can you give a rule for ascertaining the

resistance encountered by a paddle board at any part of its motion?

A.—In the common radial wheel the resistance or pressure upon any float may be ascertained as follows:—from the velocity of the centre of pressure in feet per second, subtract the velocity of the rolling circle multiplied by the cosine of the angle of inclination the paddle board makes with a vertical line; square the remainder, and multiply it by ·9715 times the area of the float in feet, which will give the tangential resistance of the float in lbs. avoirdupois. The horizontal resistance may be obtained by multiplying this quantity again by the cosine of the angle of inclination the float makes with a vertical line. In the feathering wheel the resistance may be ascertained by multiplying the velocity of the centre of pressure by the cosine of the angle of inclination, subtracting from this the velocity of the rolling circle, squaring the remainder, and multiplying it by ·9715 times the area of float in feet, which will give the resistance of the float in pounds. By thus calculating the resistance of the different floats immersed, taking the mean of these resistances and multiplying by the number of floats in the water, we readily ascertain the resistance of the wheel; and by finding what area of vertical paddle board moving at the velocity of the vessel would occasion the same resistance as all the floats immersed, and comparing this area with the sectional area of the vessel, we may find the power necessary to propel a board through the water of equal area with the cross section of the vessel, which will show the diminution of resistance consequent upon the sharpening away of the bow and stern. These relations, with many others connected with the paddle wheels, have been investigated at considerable length by Mr. Barlow in a paper published in the Philosophical Transactions, and reprinted in the Appendix to Tredgold's Treatise on the Steam Engine; but some of Mr. Barlow's deductions are erroneous, and he has vitiated nearly all his conclusions by confounding the actual with the nominal horse power.

134. *Q.*—Can you give any practical rules for proportioning paddle wheels?

A.—A common rule for the pitch of the floats is to allow one float for every foot of diameter of the wheel; but in the case of fast vessels a pitch of 2½ feet, or even less, appears preferable, as a close pitch occasions less vibration. If the floats be put too close, however, the water will not escape freely from between them; and if set too far apart, the stroke of the entering paddle will occasion an inconvenient amount of vibratory motion, and there will also be some loss of power. To find the proper area of a single float:—divide the number of actual horses power of both engines by the diameter of the wheel in feet; the quotient is the area of one paddle board in square feet proper for sea-going vessels, and the area multiplied by 0·6 will give the length of the float in feet. In very sharp vessels, which offer less resistance in passing through the water, the area of paddle board is usually one-fourth less than the above proportion, and the proper length of the float may in such case be found by multiplying the area by 0·7. In sea-going vessels about four floats are usually immersed, and in river steamers only one or two floats. There is more slip in the latter case, but there is also more engine power exerted in the propulsion of the ship, from the greater speed of engine thus rendered possible. If to permit a greater speed of the engine the floats be diminished in area instead of being raised out of the water, no appreciable accession to the speed of the vessel will be obtained; whereas there will be an increased speed of vessel if the accelerated speed of the engine be caused by diminishing the diameter of the wheels. In vessels intended to be very fast, therefore, it is expedient to make the wheels small, so as to enable the engine to work with a high velocity; and it is expedient to make such wheels of the feathering kind to obviate loss of power from oblique action. In no wheel must the rolling circle fall below the water line, else the entering and emerging floats will carry masses of water before them. The slip is usually equal to about one-fourth of the velocity of the centre of pressure in well-proportioned wheels; but it is desirable to have the slip as small as is possible consistently with the observance of other necessary conditions. The speed of the engine and also the speed of the vessel being fixed, the diameter of the rolling circle becomes at once ascertainable, and adding to this the slip, we have the diameter of the wheel.

135. *Q.*—Is the screw propeller as effectual an instrument of propulsion as the radial or feathering paddle?

A.—In all cases of deep immersion it appears to be quite as effectual as the radial paddle; but it is scarcely as effectual as the feathering paddle, with any amount of immersion, and scarcely as effectual as the common paddle in the case of light immersions.

136. *Q.*—In what way are the dimensions of the

screw propeller proportioned to the power of the engines and area of midship-section of vessel?

A.—It has been found experimentally, that when the screw is so proportioned as to make the slip about $\frac{1}{10}$th of the speed of the screw, a maximum effect is produced, the speed of the vessel being then $\frac{9}{10}$ths of what it would be if the screw were working in a solid. Find, by some of the rules previously given, what speed is due to the form and power as if the vessel were to be propelled by paddles, which speed call $\frac{9}{10}$ths, and adding $\frac{1}{10}$th for slip, we shall have the number of feet per minute to be travelled by the screw, and this number divided by the number of revolutions per minute, will give the pitch of the screw in feet. The length of screw that is found most beneficial is about $\frac{1}{6}$th of a convolution, and the diameter should be as large as it can be got. A screw with two arms, or a portion of a double-threaded screw, has been found as effectual a propeller as any other; but a screw with three blades, or a portion of a three-threaded screw, has been found to act with a more equable and regular motion. The stern post carries behind it a certain quantity of dead water which moves with the vessel, and when the screw gets into this water, which happens in the case of a double-threaded screw when the two arms of the screw are vertical, the resistance becomes nearly equal to that which would be due to the whole velocity of the screw, instead of the difference of velocity of screw and vessel; and the forward thrust at that point is so much increased thereby, that an uneasy motion is given to the hinder part of the ship, while a waste of power is at the same time occasioned. Where three blades are employed, this vibratory motion is by no means so conspicuous, as the impulse is equally divided among the three blades; and a three-bladed screw therefore appears, on the whole, to be preferable. It is very important to make the run of the ship very fine, so that the vessel may carry as small a quantity as possible of dead water behind her. As regards the form of screw, it is found that the common screw with a uniform pitch is as effectual as any: screws of an increasing pitch have been tried at various times, but without the realization of any appreciable advantage. The water thrown backwards by the screw assumes a conical form, and the obliquity and consequent loss of power will be greater when the pitch of the screw is coarse and the diameter small; so that it is expedient to have as large a diameter as possible, a quick speed, and a fine pitch.

PART II.

STRENGTHS, CONSTRUCTIVE DETAILS, AND MANAGEMENT.

137. *Q.*—In what way are the strengths of the different parts of a steam engine determined?

A.—By a reference to the cohesion of iron, or its power to resist a tensile, twisting, breaking, or crushing force—which has been fixed by numerous experiments, and by the computation of the amount of strain to which the several parts are subjected, to the end that the quantity of material may be made proportionate to the strain. The breaking strain of a bar of malleable iron, of medium quality, one inch square, when pulled in the direction of its length, is from 50,000 to 60,000 lbs., and of a bar of cast-iron from 17,000 to 20,000 lbs.; but a much smaller strain than this will damage the structure of the iron, and finally break it, if the strain be permitted to act without interruption. The tensile strain to which a bar of malleable iron an inch square may be subjected, without permanently deranging its structure, is 17,800 lbs.; and in the case of cast-iron 15,300 lbs.: but it would not be safe to apply such a strain in practice, as there are so many different qualities of iron, and so many irregularities of structure even in iron of the same quality, from unequal conversion, imperfect welds, and other circumstances, that it would be unwarrantable to reckon the maximum strength as at all times subsisting. The greatest strain to which malleable iron should be subjected in machinery is 4000 lbs. on the square inch of section, though in locomotive boilers this strain is sometimes exceeded. In the case of beams subjected to a breaking force, the strength with any given cohesion of the material will be proportional to the breadth, multiplied by the square of the depth; and in the case of revolving shafts exposed to a twisting strain, the strength with any given cohesive power of the material will be as the cube of the diameter. If the force acting at the end of an engine beam be taken at 18 lbs. per circular inch of the piston, then the force acting at the middle will be 36 lbs. per circular inch of the piston, and the proper strength of the beam at the centre will be found by the following rule:—divide the weight in lbs. acting at the centre by 250, and multiply the quotient by the distance between the extreme centres. To find the depth, the breadth being given: divide this product by the breadth in inches, and extract the square root of the quotient, which is the depth. The depth of an engine beam at the ends is usually made one-third of the depth at the centre, which is equal to the diameter of the cylinder in the case of low pressure engines, while the length is made equal to three times the length of the stroke, and the mean thickness $\frac{1}{108}$th of the length—the width of the edge bead being about three times the thickness of the web. In low pressure engines the diameter of the end studs of the engine beam are usually made $\frac{1}{9}$th of the diameter of the cylinder when of cast-iron, and $\frac{1}{10}$th when of wrought-iron, which gives a load with low steam of about 500 lbs. per circular inch of transverse section; but a larger size is preferable, as with large bearings the brasses do not wear so rapidly and the straps are not so likely to be burst by the bearings becoming oval. To find the proper size of a cast-iron gudgeon adapted to sustain any given weight: multiply the weight in lbs. by the intended length of bearing expressed in terms of the diameter; divide the product by 500, and extract the square root of the quotient, which is the diameter in inches. Experiments upon the force requisite to twist off cast-iron necks show that if the cube of the diameter of neck in inches be multiplied by 880, the product will be the force of torsion which will twist them off when acting at 6 inches radius. To find the diameter of a cast-iron fly wheel shaft: multiply the square of the diameter of the cylinder in inches by the length of the crank in inches, and extract the cube root of the product, which multiply by 0·3025, and the result will be the diameter of the shaft in inches at the smallest part when of cast-iron. To find the diameter of the paddle shaft of a steam vessel when of wrought-iron: multiply the square of the diameter of cylinder in inches by the length of the crank in inches, and extract the cube root of the product, which multiply by 0·242; the result is the diameter of the shaft in inches at the smallest part when of malleable iron. The diameter of the crank pin is usually made $\frac{1}{6}$th of the diameter of the cylinder when of cast-iron, and $\frac{1}{7}$th of the diameter of the cylinder when of malleable iron. The diameter of the piston rod is usually made $\frac{1}{10}$th of the diameter of the cylinder, or the sectional area of the piston rod is $\frac{1}{100}$th of the area of the cylinder. The sectional area of the main links in land beam engines is $\frac{1}{113}$th of the area of the cylinder, and the length of the main links is usually half the length of the stroke. In land engines the connecting rod is usually of cast-iron, with a cruciform section: the breadth across the arms of the cross is about $\frac{1}{20}$th of the length of the rod, the sectional area at the centre $\frac{1}{28}$th of the area of the cylinder, and at the ends $\frac{1}{35}$th of the area of the cylinder: the length of the rod is usually $3\frac{1}{2}$ times the length of the stroke. To find the proper dimensions for the teeth of a cast-iron wheel: multiply the diameter of the pitch circle in feet by the number of revolutions to be made per minute, and reserve the product for a divisor; multiply the number of *actual* horses power to be transmitted by 240, and divide the product by the above divisor, which will give the strength. If the pitch be given to find the breadth, divide the above strength by the square of the pitch in inches ; or if the breadth be given, then to find the pitch, divide the strength by the breadth in inches, and extract the square root of the quotient, which is the proper pitch in inches. The length of the teeth is usually about $\frac{5}{8}$ths of the pitch. Pinions to work satisfactorily should not have less than 30 or 40 teeth, and where the speed exceeds 220 feet in the minute, the teeth of the larger wheel should be of wood, made a little thicker to keep the strength unimpaired. These rules are for the most part applicable only in the case of condensing engines working with steam of a few pounds pressure above the atmosphere ; whereas in many modern engines, and especially in marine engines since the introduction of tubular boilers, the pressure of the steam has been so much increased as to make the force urging the piston twice greater than formerly.

138. *Q.*—Cannot you give some rules of strength which will be applicable whatever pressure may be employed ?

A.—In the rules already given, the pressure may be reckoned at from 18 to 20 lbs. upon every square inch of the piston ; and if the pressure upon every square inch of the piston be made twice greater, the dimensions must just be those proper for an engine of twice the area of piston. It will not be difficult, however, to introduce the pressure into the rules as an element of the computation, whereby the result will be applicable both to high and low pressure engines.

The method of computation will then be as follows: to find the dimensions of a malleable iron paddle shaft, so that the strain shall not exceed $\frac{5}{6}$ths of the elastic force, or $\frac{5}{6}$ths of the force iron is capable of withstanding without permanent derangement of structure, which in tensile strains is 17,800 lbs. per square inch: multiply 0·08264 times the pressure in lbs. per square inch on the piston by the square of the diameter of the cylinder in inches, and the length of the crank in inches, and extract the cube root of the product, which will be the diameter of the paddle shaft journal in inches when of malleable iron, whatever the pressure of the steam may be. The length of the paddle shaft journal should be $1\frac{1}{4}$ times the diameter; and the diameter of the part where the crank is put on is often made equal to the diameter over the collars of the journal or bearing. To find the exterior diameter of the large eye of the crank when of malleable iron:—to 1·561 times the pressure of the steam upon the piston in lbs. per square inch, multiplied by the square of the length of the crank in inches, add 0·00494 times the square of the diameter of the cylinder in inches, multiplied by the square of the number of lbs. pressure per square inch on the piston; extract the square root of this quantity; divide the result by 75·59 times the square root of the length of the crank in inches, and multiply the quotient by the diameter of the cylinder in inches; square the product, and extract the cube root of the square, to which add the diameter of the hole for the reception of the shaft, and the result will be the exterior diameter of the large eye of the crank when of malleable iron. The diameter of the small eye of the crank may be found by adding to the diameter of the crank pin 0·02521 times the square root of the pressure on the piston in lbs. per square inch, multiplied by the diameter of the cylinder in inches; the diameter of the crank pin may be found by multiplying 0·02836 times the square root of the pressure on the piston in lbs. per square inch by the diameter of the cylinder in inches. The length of the pin is usually about $\frac{9}{8}$th times its diameter, and the strain if all thrown upon the end of the pin will be equal to the elastic force; but in ordinary working, the strain will only be equal to $\frac{1}{3}$d of the elastic force. The thickness of the web of the crank, supposing it to be continued to the centre of the shaft, would at that point be represented by the following rule:—to 1·561 times the square of the length of the crank in inches, add 0·00494 times the square of the diameter of the cylinder in inches, multiplied by the pressure on the piston in lbs. per square inch; extract the square root of the sum, which multiply by the diameter of the cylinder squared in inches, and by the pressure on the piston in lbs. per square inch; divide the product by 9000, and extract the cube root of the quotient, which will be the proper thickness of the web of the crank when of malleable iron, supposing the web to be continued to the centre of the shaft. The thickness of the web at the crank pin centre, supposing it to be continued thither, would be 0·022 times the square root of the pressure on the piston in lbs. per square inch, multiplied by the diameter of the cylinder. The breadth of the web of the crank of the shaft centre should be twice the thickness, and at the pin centre $1\frac{1}{2}$

times the thickness of the web; the length of the large eye of the crank should be equal to the diameter of the shaft, and of the small eye 0·0375 times the square root of the pressure on the piston in lbs. per square inch, multiplied by the diameter of the cylinder. The diameter of the piston rod may be found by multiplying the diameter of the cylinder in inches by the square root of the pressure on the piston in lbs. per square inch, and dividing by 50, which makes the strain $\frac{1}{7}$th of the elastic force. The diameter of the connecting rod at the ends, may be found by multiplying 0·019 times the square root of the pressure on the piston in lbs. per square inch by the diameter of the cylinder in inches; and the diameter of the connecting rod in the middle may be found by the following rule:—to 0·0035 times the length of the connecting rod in inches, add 1, and multiply the sum by 0·019 times the square root of the pressure on the piston in lbs. per square inch, multiplied by the diameter of the cylinder in inches. The diameter of the cylinder side rods at the ends may be found by multiplying 0·0129 times the square root of the pressure on the piston in lbs. per square inch by the diameter of the cylinder; and the diameter of the cylinder side rods at the middle is found by the following rule:—to 0·0035 times the length of the rod in inches add 1, and multiply the sum by 0·0129 times the square root of the pressure on the piston in lbs. per square inch, multiplied by the diameter of the cylinder in inches; the product is the diameter of each side rod at the centre in inches. The strain upon the connecting rod and side rods is by these rules equal to $\frac{1}{8}$th of the elastic force. If the length of the cross head be taken at 1·4 times the diameter of the cylinder, the dimensions of the cross head will be as follows:—the exterior diameter of the eye in the cross head for the reception of the piston rod, will be equal to the diameter of the hole, plus 0·02827 times the cube root of the pressure on the piston in lbs. per square inch, multiplied by the diameter of the cylinder in inches; and the depth of the eye will be 0·0979 times the cube root of the pressure on the piston in lbs. per square inch, multiplied by the diameter of the cylinder in inches. The diameter of each cross head journal will be 0·01716 times the square root of the pressure on the piston in lbs. per square inch, multiplied by the diameter of the cylinder in inches—the length of the journal being $\frac{9}{8}$ths its diameter. The thickness of the web at centre will be 0·0245 times the cube root of the pressure on the piston in lbs. per square inch, multiplied by the diameter of the cylinder in inches, and the depth of web at centre will be 0·09178 times the cube root of the pressure on the piston in lbs. per square inch, multiplied by the diameter of the cylinder in inches. The thickness of the web at journal will be 0·0122 times the square root of the pressure on the piston in lbs. per square inch, multiplied by the diameter of the cylinder in inches, and the depth of the web at journal will be 0·0203 times the square root of the pressure upon the piston in lbs. per square inch, multiplied by the diameter of the cylinder in inches. In these rules for the cross head, the strain upon the web is $\frac{1}{2.225}$ times the elastic force; the strain upon the journal in ordinary working is $\frac{1}{2.33}$ times the elastic force; and if the outer ends of the journals are the only bearing points, the strain is $\frac{1}{1 \cdot 165}$ times the elastic force, which is very little in excess of the elastic force. The diameter of the main centre may be found by multiplying 0·0367 times the square of the pressure upon the piston in lbs. per square inch, by the diameter of the cylinder in inches, which will give the diameter of the main centre journal when of malleable iron, and the length of the main centre journal should be $1\frac{1}{2}$ times its diameter; the strain upon the main centre journal in ordinary working will be about $\frac{1}{2}$ the elastic force. In oscillating engines the diameter of the trunnions is regulated by the diameter of the pipes, and the thickness of the trunnions is about the same as the thickness of the cylinder. It is obvious, from the varying proportions subsisting in the different parts of the engine between the strain and the elastic force, that in engines proportioned by these rules—which represent nevertheless the average practice of the best constructors—some of the parts must possess a considerable excess of strength over other parts, and it appears expedient that this disparity should be diminished, which may best be done by increasing the strength of the parts which are weakest; inasmuch as the frequent fracture of some of the parts shows that the dimensions at present adopted for those parts are scarcely sufficient, unless the iron of which they are made is of the best quality. At the same time it is quite certain, that engines proportioned by these rules will work satisfactorily where good materials are employed; but it is important to know in what parts good materials and large dimensions are the most indispensable. The depth of gibs and cutters for attaching the piston rod to the cross head, is 0·0358 times the cube root of the pressure of the steam on the piston in lbs. per square inch, multiplied by the diameter of the cylinder; and the thickness of the gibs and cutters is 0·007 times the cube root of the pressure on the piston in lbs. per square inch, multiplied by the diameter of the cylinder. The depth of the cutter through the piston is 0·017 times the square root of the pressure on the piston in lbs. per square inch, multiplied by the diameter of the cylinder in inches; and the thickness of the cutter through the piston is 0·007 times the square root of the pressure on the piston in lbs. per square inch, multiplied by the diameter of the cylinder. The whole of the answers to these rules are given in inches.

139. *Q.*—How do you determine the strength of boilers?

A.—The iron of boilers, like the iron of machines or structures, is capable of withstanding a tensile strain of from 50,000 to 60,000 lbs. upon every square inch of section; but it will only bear a third of this strain without permanent derangement of structure, and it does not appear expedient in any boiler to let the strain exceed 4000 lbs. upon the square inch of sectional area of metal; and 3000 lbs. on the square inch of section, is a preferable proportion. The question of the strength of boilers was investigated very elaborately a few years ago, by a Committee of the Franklin Institute, in America, and it was found that the tenacity of boiler plate increased with the temperature up to 550°, at which point the tenacity began to diminish. At 32° the cohesive force of a square inch of section was 56,000 lbs.; at 570° it was 66,500 lbs; at 720°, 55,000 lbs.; at 1050°, 32,000 lbs.; at 1240°, 22,000 lbs.; and at 1317°, 9000 lbs. Copper follows a different law, and appears to be diminished in strength by every addition to the temperature. At 32° the cohesion of copper was found to be 32,800 lbs. per square inch of section, which exceeds the cohesive force at any higher temperature, and the square of the diminution of strength seems to keep pace with the cube of the increased temperature. Strips of iron cut in the direction of the fibre were found to be about 6 per cent. stronger than when cut across the grain. Repeated piling and welding was found to increase the tenacity of the iron, but the result of welding together different kinds of iron was not found to be favorable. The accidental overheating of a boiler was found to reduce the ultimate or maximum strength of the plates from 65,000 lbs. to 45,000 lbs. per square inch of section, and riveting the plates was found to occasion a diminution in their strength to the extent of one-third. In some locomotive boilers which are worked with a pressure of 80 lbs. upon the square inch, the thickness of the plates is only $\frac{5}{16}$ths of an inch, while the barrel of the boiler is 39 inches in diameter. It will require a length of 3·2 inches of the boiler when the plates are $\frac{5}{16}$ths thick to make up a sectional area of one square inch, and the separating force will be 39 times 3·2 multiplied by 80, which makes the separating force 9984 lbs. sustained by two square inches of sectional area,—one on each side; or the strain is 4992 lbs. per square inch of sectional area, which is a greater strain than is advisable. The accession of strength derived from the boiler ends is not here taken into account; but neither is the weakening effect counted that is caused by the rivets, which certainly would not be less in amount. The proper thickness for cylindrical boilers or other cylindrical vessels, whether of cast or wrought iron, exposed to an internal pressure, may be found by the following rule:—multiply 2·54 times the internal diameter of the cylinder in inches by the greatest pressure within the cylinder per circular inch, and divide by the tensile force that the metal will bear without permanent derangement of structure, which for malleable iron is 17,800 lbs., and for cast-iron 15,300 lbs. per square inch of section; the result is the thickness in inches. Where the sides of the boiler are flat instead of being cylindrical, a sufficient number of stays must be introduced to withstand the pressure, and it is expedient not to let the strain upon these stays be more than 3000 lbs. per square inch of section, as the strength of internal stays in boilers is generally soon diminished by corrosion. It is expedient also that the stays should be small and numerous rather than large and few in number, as when large stays are employed it is difficult to keep them tight at the ends, and oxidation of the shell follows from leakage at the ends of the stays. A strain at all approaching that upon locomotive boilers would be very unsafe in the case of marine boilers, on account of the corrosion, both internal and external, to which marine boilers are subject. All boilers should be proved, when new, to three or four times the pressure they are intended to bear, and they should be proved occasionally by the hand pump when in use, to detect any weakness which corrosion may have occasioned.

140. *Q.*—What is the cause of the rapid corrosion of marine boilers?

A.—Marine boilers are corroded externally in the region of the steam chest by the dripping of water from the deck; the bottom of the boiler is corroded by the action of the bilge water, and the ash pits by the practice of quenching the ashes with salt water. These sources of injury, however, admit of easy remedy: the top of the boiler may be preserved from external corrosion by covering it with felt, upon which is laid sheet lead soldered at every joint so as to be impenetrable to water; the ash pits may be shielded by guard plates, which are plates fitting into the ash pits and attached to the boiler by a few bolts, so that when worn they may be removed and new ones substituted, whereby any wear upon the boiler in that part will be prevented; and there will be very little wear upon the bottom of a boiler if it be imbedded in mastic cement laid upon a suitable platform. The greatest part of the corrosion of a boiler, however, takes place in the inside of the steam chest, and the origin of this corrosion is one of the obscurest subjects in the whole range of engineering. It cannot be from the chemical action of the salt water upon the iron, for the flues and other parts of the boiler beneath the water suffer very little from corrosion; and in steam vessels provided with Hall's condensers, which supply the boiler with fresh water, scarcely any increased durability of the boiler has been experienced. Nevertheless, marine boilers seldom last more than for 4 or 5 years, whereas land boilers made of the same quality of iron often last 18 or 20 years, and it does not appear probable that land boilers would last a very much shorter time if salt water were used in them. The thin film of scale spread over the parts of a marine boiler situated beneath the water, effectually protects them from corrosion; and when the other parts are completely worn out, the flues generally remain so perfect that the hammer marks upon them are as conspicuous as at their first formation. The operation of the steam in corroding the interior of the boiler is most capricious—the parts which are most rapidly worn away in one boiler being untouched in another; and in some cases one side of a steam chest will be very much wasted away while the opposite side remains uninjured. Sometimes the iron exfoliates in the shape of a black oxide which comes away in flakes like the leaves of a book, while in other cases the iron appears as if eaten away by a strong acid which had a solvent action upon it. The application of felt to the outside of a boiler, has in several cases been found to accelerate sensibly its internal corrosion; boilers in which there is a large accumulation of scale appear to be more corroded than where there is no such deposit, and where the funnel passes through the steam chest, the iron of the steam chest is invariably much more corroded than where the funnel does not pass through it. These facts appear to indicate that the internal corrosion of marine boilers is attributable chiefly to the existence of surcharged steam within them, which is steam to which an additional quantity of heat has been communicated subsequently to its generation, so that its temperature is greater than is due to its elastic force; and on this hypothesis the observed facts relative to corrosion become easily explicable. Felt, applied to the outside of a boiler, may accelerate its internal corrosion by keeping the steam in a surcharged state, when by the dispersion of a part of the heat it would cease to be in that state. Boilers in which there is a large accumulation of scale must have worked with the water very salt, which necessarily produces surcharged steam; for the temperature of steam cannot be less than that of the water from which it is generated, and inasmuch as the boiling point of water, under any given pressure, rises with the saltness of the water, the temperature of the steam must rise with the saltness of the water, the pressure remaining the same; or in other words, the steam must have a higher temperature than is due to its elastic force, or be in the state of surcharged steam. The circumstance of the chimney flue passing through the steam will manifestly surcharge the steam with heat, so that all the circumstances which are found to accelerate corrosion, are, it appears, such as would also induce the formation of surcharged steam. Besides, the natural effect of surcharged steam is to oxidate the iron with which it is in contact, as is illustrated by the familiar process for making hydrogen gas by sending steam through a red-hot tube filled with pieces of iron; and although the action of the surcharged steam in a boiler is necessarily very much weaker than where the iron is red hot, it manifestly must have *some* oxidizing effect, and the amount of corrosion produced may be very material where the action is perpetual. Boilers with a large extent of heating surface, or with descending flues circulating through the cooler water in the bottom of the boiler before ascending the chimney, will be less corroded internally than boilers in which a large quantity of the heat passes away in the smoke; and the corrosion of the boiler will be diminished if the interior of any flue passing through the steam be coated with fire-brick, so as to prevent the transmission of the heat in that situation. The best practice, however, appears to consist in the transmission of the smoke through a suitable orifice below the water level, so as to supersede the necessity of carrying any flue through the steam at all; or a column of water may be carried round the chimney, into which as much of the feed water may be introduced as the heat of the chimney is capable of raising to the boiling point, as under this limitation the presence of feed water around the chimney in the steam chest will fail to condense the steam.

141. *Q.*—Will you explain the course of procedure in the construction and setting of wagon boilers?

A.—Most boilers are made of plates three-eighths of an inch thick, and the rivets are from three-eighths to three-fourths of an inch in diameter. In the bottom and sides of a wagon boiler, the heads of the rivets, or the ends formed before the rivets are inserted, should be large, and placed next the fire, or on the outside; whereas on the top of the boiler the heads should be on the inside. The rivets should be placed about two inches distant from centre to centre, and the centre of the row of rivets should be about one inch from the edge of the plate. The edges of the plate should be truly cut, both inside and outside, and after the parts of the boiler have been riveted together, the edges of the plate should be set up or caulked with a blunt chisel about a quarter of an inch thick in the point, and struck by a hammer of about 3 or 4 lbs. weight—one man holding the caulking tool while another strikes. The boiler should then be filled with water and caulked afresh in any leaky part; all the joints above the water should next be painted with a solution of sal ammoniac in urine, and so soon as the seams are well rusted they should be dried with a gentle fire, and then be painted over with a thin putty formed of whiting and linseed oil—the heat being continued until the putty becomes so hard that it cannot be readily scratched with the nail, and care must be taken neither to burn the putty nor to discontinue the fire until it has become quite dry. In building the brick-work for the setting of the boiler, the parts upon which the heat acts with most intensity is to be built with clay instead of mortar; but mortar is to be used on the outside of the work. Old bars of flat iron may be laid under the boiler chime to prevent that part of the boiler from being burned out, and bars of iron should also run through the brick-work to prevent it from splitting. The top of the boiler is to be covered with brick-work laid in the best lime; and if the lime be not of the hydraulic kind, it should be mixed with Dutch terrass to make it impenetrable to water. The top of the boiler should be well plastered with this lime, which will greatly conduce to the tightness of the seams. Openings into the flues must be left in convenient situations to enable the flues to be swept out when required, and these openings may be closed with cast-iron doors jointed with clay or mortar, which may be easily removed when required. Adjacent to the chimney a slit must be left in the top of the flue with a groove in the brick-work to enable a sliding door or damper to be fixed in that situation, which by being lowered into the flue will obstruct the passage of the smoke and moderate the draught, whereby the chimney will be prevented from drawing the flame into it before the heat has acted sufficiently upon the boiler.

142. *Q.*—Are marine constructed in the same way as land boilers?

A.—There is very little difference in the two cases: the whole of the shells of marine boilers, however, should be double riveted with rivets $\frac{11}{16}$th of an inch in diameter, and $2\frac{3}{8}$th inches from centre to centre, the weakening effect of double riveting being much less than that of single riveting. The furnaces above the line of bars should be of the best Lowmoor, Bowling, or Staffordshire scrap plates, and the portion of each furnace above the bars should consist only of three plates, one for the top and one for each side, the lower seam of the side plates being situated beneath the level of the bars, so as not to be exposed to the heat of the furnace. The tube plates of tubular boilers should be of the best Lowmoor or Bowling iron, seven-eighths to one inch thick: the shells should be of the best Staffordshire or Thornycroft S Crown iron, seven-sixteenths of an inch thick. Angle iron should not be used in the construction of boilers, as in the manufacture it becomes reedy, and is apt to split up in the direction of its length: it is much the safer practice to bend the plates at the corners of the boiler, but this must be carefully done, without introducing any more sharp bends than can be avoided, and plates which require to be bent much should be of Lowmoor iron. It will usually be found expedient to introduce a ring of angle iron around the furnace mouths, though it is discarded in the other parts of the boiler; but it should be used as sparingly as possible, and any that is used should be of the best quality. The whole of the plates of a boiler should have the holes for the rivets punched, and the edges cut straight, by means of self-acting machinery, in which a travelling table carries forward the plate with an equal progression every stroke of the punch or shears ; and machinery of this kind is now extensively employed. The practice of forcing the parts of boilers together with violence, by means of screw-jacks, and drifts through the holes, should not be permitted; as a great strain may thus be thrown upon the rivets, even when there is no steam in the boiler. All rivets should be of the best Lowmoor iron. The work should be caulked both within and without wherever it is accessible; but in the more confined situations within the flues, the caulking will in many cases have to be done with the hand or chipping instead of the heavy hammer previously prescribed. In the setting of marine boilers care must be taken that no copper bolts or nails project above the platform upon which they rest, and also that no projecting copper bolts in the sides of the ship touch the boiler, as the galvanic action in such a case would probably soon wear the points of contact into holes. The platform may consist of 3-inch planking laid across the keelsons nailed with iron nails, the heads of which are well punched down, and caulked and puttied like a deck. The surface may then be painted over with thin putty, and fore and aft boards of half the thickness may then be laid down and nailed securely with iron nails, having the heads well punched down. This platform must then be covered thinly and evenly with mastic cement, and the boiler be set down upon it, and the cement must be caulked beneath the boiler by means of wooden caulking tools, so as completely to fill every vacuity. Coomings of wood sloped on the top must next be set round the boiler, and the space between the coomings and the boiler must be caulked full of cement, and be smoothed off on the top to the slope of the coomings, so as to throw off any water that might be disposed to enter between the coomings and the boiler. Mastic cement proper for the setting of boilers is sold in many places ready made; Hamelin's mastic is compounded as follows:—To any given weight of sand or pulverized earthenware add two-thirds such given weight of powdered Bath, Portland, or other similar stone, and to every five hundred and sixty pounds weight of the mixture add forty pounds weight of litharge, two pounds of powdered glass or flint, one pound of minium, and two pounds of gray oxide of lead; pass the mixture through a sieve, and keep it in a powder for use. When wanted for use, a sufficient quantity of the powder is mixed with some vegetable oil upon a board or in a trough in the manner of mortar, in the proportion of six hundred and five pounds of the powder to five gallons of linseed, walnut, or pink oil, and the mixture is stirred and trodden upon until it assumes the appearance of moistened sand, when it is ready for use. The cement should be used on the same day the oil is added, else it will set into a solid mass.

143. *Q.*—Will you state any other constructive details in connection with marine boilers which occur to you as deserving of enumeration?

A.—It will not be possible to state them in a very systematic manner by adopting this course, but perhaps it may be of most importance to dispatch them summarily. It has already been stated that furnace bars should not much exceed 6 ft. in length, as it is difficult to manage long furnaces; but it is a frequent practice to make furnaces long and narrow, the consequence of which is, that it is impossible to fire them effectually at the after end, especially upon long voyages and in stormy weather, and air escapes into the flues at the after end of the bars, whereby the efficacy of the boiler is diminished. Where the bars are very long, it will generally be found that an increased supply of steam and a diminished consumption of coal will be the consequence of shortening them; and the bars should always lie with a considerable inclination, to facilitate the distribution of the fuel over the after part of the furnace. When there are two lengths of bars in the furnace, it is expedient to make the central cross-bar for bearing up the ends double, and to leave a space between the ends of the bars so that the ashes may fall through between them. The space thus left enables the bars to expand without injury on the application of heat; whereas, without some such provision, the bars are very liable to get burned out by bending up in the centre, or at the ends—as they must do if the elongation of the bars on the application of heat be prevented; and this must be the effect of permitting the spaces at the ends of the bars to be filled up with ashes. At each end of each bed of bars it is expedient to leave a space which the ashes cannot fill up so as to cause the bars to jam; and care must be taken that the heels of the bars do not come against any of the furnace bearers, whereby the room left at the end of the bars to permit the expansion would be rendered of no avail. The furnace bridges of marine boilers are either made of fire-brick or of plate iron containing water: in the case of water bridges, the top part of the bridge should be made with a large amount of slant, so as to enable the steam to escape freely; but, notwithstanding this precaution, the plates of water bridges are apt to crack at the bend, so that fire-brick bridges appear on the whole to be preferable. In shallow furnaces the bridges often come too near the furnace top to enable a man to pass over them, and it will save expense if, in such bridges, the upper portion is constructed of two or three fire-blocks, which may be lifted off where a person requires to enter the flues to sweep or repair them, whereby the perpetual demolition and reconstruction of the upper part of the bridge will be prevented. The furnaces are shallowest in boilers with a double tier of furnaces one above the other; but such boilers are now little used: the steam of the lower furnace appears to be condensed to a certain extent in coming into contact with the iron of the superior ash-pit, and it is difficult, in such shallow furnaces as the plan involves, to give the bars a sufficient inclination to enable the fuel to pass on to the after end. The flue of flue boilers generally contracts in area as it approaches the chimney; and it is a common practice to place a hanging bridge, consisting of a plate of iron descending a certain distance into the flue, at that part of the flue where it enters the chimney, whereby the stratum of hot air which occupies the highest part of the flue is kept in protracted contact with the boiler, and the cooler air occupying the lower part of the flue is that which alone escapes. The practice of introducing a hanging bridge is a beneficial one in the case of some boilers, but is not applicable universally, as boilers with a small calorimeter cannot be further contracted in the flue without a diminution in their evaporating power. In tubular boilers a hanging bridge is not applicable; but in some cases a perforated plate is placed against the ends of the tubes, which, by suitable connections, is made to operate as a sliding damper, which partially or totally closes up the end of every tube, and, at other times, a damper constructed in the manner of a Venetian blind is employed in the same situation. These varieties of damper, however, have only yet been used in locomotive boilers, though applicable to tubular boilers of every description. It is an advantage that there should be a division between the tubes pertaining to each furnace of a marine tubular boiler, so that the smoke of each furnace may be kept apart from the smoke of the furnace adjoining it until the smoke of both enters the chimney, as by this arrangement a furnace only will be rendered inoperative in cleaning the fires instead of a boiler, and the tubes belonging to one furnace may be swept if necessary at sea without interfering injuriously with the action of the rest: the connection between the tubes of the inactive furnace and the chimney would obviously require to be closed in such a case, either by a damper or otherwise. The sides of the internal furnaces or flues in all boilers should be so constructed that the steam may readily escape from their surfaces, with which view it is expedient to make the bottom of the flue somewhat wider than the top, or slightly conical in the cross section, and the upper plates should always be overlapped by the plates beneath, so that the steam cannot be retained in the overlap, but escapes as soon as it is generated. If the sides of the furnace be made high and perfectly vertical, they will speedily be buckled and cracked by the heat, as a film of steam in such a case will remain in contact with the iron, which will prevent the access of the water, and the iron of the boiler will be injured by the high temperature it must in that case acquire. To moderate the intensity of the heat acting upon the furnace sides, it is expedient to bring the outside fire-bars into close contact with the sides of the furnace, so as to prevent the entrance of air through the fire in that situation, by which the intensity of the heat would be increased. The tube plate nearest the furnace in tubular boilers should also be so inclined as to facilitate the escape of the steam; and the short bent plate or flange of the tube plate, connecting the tube plate with the top of the furnace, should be made with a gradual bend, as if the bend be sudden the iron will be apt to crack or burn away from the concretion of salt. Where the furnace mouths are contracted by bending in the sides and top of the furnace, as is the general practice, the bends should be gradual, as salt is apt to accumulate in the pockets made by a sudden bend, and the plates will then burn into holes. It is very expedient that sufficient space should be left between the furnace and the tubes in all tubular boilers to permit a boy to go in to clear away any scale that may have formed, and to hold on the rivets in the event of repair being wanted; and it is also expedient that a vertical row of tubes should be left out opposite to each water space to allow the ascent of the steam and the descent of the water, as it has been found that the removal of the tubes in that position, even in a boiler with deficient heating surface, has increased the production of steam, and diminished the consumption of fuel. The tubes should all be kept in the same vertical line, so as to permit the introduction of an instrument to scrape them, but they should be zig-zagged in the horizontal line, whereby a greater strength of metal will be obtained around the holes in the tube plates.

144. *Q.*—In what manner is the tubing and staying of boilers performed?

A.—The tubes of marine boilers are generally iron tubes, 3 inches in diameter, and between 6 and 7 ft. long, but sometimes brass tubes of similar dimensions are employed. When brass tubes are employed, the use of ferules driven into the ends of the tubes appears to be indispensable to keep them tight; but when the tubes are of malleable iron, of the thickness of Russell's boiler tubes, they may be made tight merely by driving them firmly into the tube plates. The holes in the tube plate next the front of the boiler are made about one-sixteenth of an inch larger in diameter than the holes in the other tube plate, and the holes upon the outer surfaces of both tube plates are slightly countersunk. One end of each tube is enlarged about a sixteenth of an inch to fit the enlarged holes in one of the plates; both ends may then be turned slightly in the lathe to make them smooth and uniform in diameter and length, and the whole of the tubes are then to be all driven through both tube plates from the front of the boiler—the precaution, however, being taken to drive them in gently at first with a light hand-hammer, until the whole of the tubes have been inserted to an equal depth, and then they may be driven up by degrees with a heavy hammer, whereby any distortion of the holes from unequal driving will be prevented. Finally, the ends of the tubes should be riveted up so as to fill the countersink; the tubes should be left a little longer than the distance between the outer surfaces of the tube plates, so that the countersink at the ends may be filled by staving up the end of the tube rather than by riveting it over, and the staving will be best accomplished by means of a mandril with a collar upon it, which is driven into the tube so that the collar rests upon the end of the tube to be riveted. It appears expedient in all cases where ferules are not used, that some of the tubes should be screwed at the ends, so as to serve as stays if the riveting at the tube ends happens to be burned away, and also to act as abutments to the riveted tube. To prevent leakage through the thread of the screw, if the plan of screwing some of the tubes be adopted, it will be expedient to let the screwed ends project about half an inch, and to put thin nuts upon them with a little white lead interposed. If the tubes are long, their expansion when the boiler is being blown off will be apt to start them at the ends, unless very securely fixed, and it is impossible to prevent brass tubes of large diameter and proportionate length from being started at the ends, even when secured by ferules ; but the brass tubes commonly employed are so small as to be susceptible of sufficient compression endways by the adhesion due to the ferules to compensate for the expansion, whereby they are prevented from starting at the ends. In some of the early marine boilers fitted with brass tubes, a galvanic action at the ends of the tubes was found to take place, and the iron of the tube plates was wasted away in consequence, with rapidity ; but further experience proved the injury to be attributable chiefly to imperfect fitting, whereby a leakage was caused that induced oxidation: and where the tubes were well fitted, any injurious action at the ends of the tubes was found to cease. When the pressure of steam within the boiler is considerable, the boiler must be very securely stayed: the top and bottom, and also the sides of the boiler must be stayed to one another, and it will not be sufficient to stay the top of the boiler to the top of the furnaces, and the bottom of the boiler to the bottom of the furnaces; for if a furnace changed its form, as it would be likely to do in such circumstances, the stays connecting it to the top and bottom of the boiler would be of very little utility in preventing the boiler from bursting. If the pressure of steam be 20 lbs. on the square inch, which is a very common pressure in tubular boilers, there will be a pressure of 2880 lbs. on every square foot of flat surface, so that if the strain upon the stays is not to exceed 3000 lbs. on the square inch of section, there must be nearly a square inch of sectional area of stay for every square foot of flat surface on the top and bottom, sides, and ends of the boiler. This very much exceeds the proportion usually adopted, and in scarcely any instance are boilers stayed sufficiently to be safe when the shell is composed of flat surfaces. The furnaces should be stayed together with bolts of the best scrap iron $1\frac{1}{4}$ inch in diameter, tapped through both plates of the water space with thin nuts in each furnace; and it is expedient to make the row of stays running horizontally near the level of the bars, sufficiently low to come beneath the top of the bars, so as to be shielded from the action of the fire, with which view they should follow the inclination of the bars. The row of stays between the level of the bars and the top of the furnace should be as near the top of the furnace as will consist with the functions they have to perform, so as to be removed as far as possible from the action of the heat; and to support the furnace, top cross-bars may either be adopted, to which the top is secured with bolts as in the case of locomotives, or stays tapped into the furnace top with a thin nut beneath may be carried to the top of the boiler; but very little dependance can be put in such stays as stays for keeping down the top of the boiler, and the top of the boiler must therefore be stayed nearly as much as if the stays connecting it with the furnace crowns did not exist. The large rivets passing through thimbles, sometimes used as stays for water spaces or boiler shells, are objectionable; as from the great amount of hammering such rivets have to receive to form the heads, the iron becomes crystalline, so that the heads are liable to come off, and indeed sometimes fly off in the act of being formed. If such a fracture occurs between the boilers after they are seated in their place, or in any position not accessible from the outside, it will in general be necessary to empty the faulty boiler, and repair the defect from the inside. The stays where the sides of the boiler are flat and the pressure of the steam is from 20 to 30 lbs., should be pitched about a foot or 18 inches asunder; and in the wake of the tubes where stays cannot be carried across to connect the boiler sides, angle iron ribs, like the ribs of a ship, should be riveted to the interior of the boiler, and stays of greater strength than the rest should pass across above and below the tubes, to which the angle irons would communicate the strain. The whole of the long stays within a boiler should be firmly riveted to the shell, as if built with and forming a part of it; as by the common method of fixing them in by means of cutters, the decay or accidental detachment of a pin or cutter may endanger the safety of the boiler. Wherever a large perforation in the shell of any circular boiler occurs, a sufficient number of stays should be put across it to maintain the original strength; and where stays are intercepted by the root of the funnel, short stays in continuation of them should be placed inside. The flues of all flue boilers diminish in their calorimeter as they approach the chimney: some very satisfactory boilers have been made by allowing a proportion of 0·6 of a square foot of fire grate per nominal horse power, and making the sectional area of the flue at the largest part $\frac{1}{7}$th of the area of fire grate, and at the smallest part where it enters the chimney $\frac{1}{11}$th of the area of the fire grate; but in some of the boilers proportioned on this plan the maximum sectional area is only $\frac{1}{7 \cdot 5}$ or $\frac{1}{8 \cdot 5}$, according to the purposes of the boiler. These proportions are retained whether the boiler is flue or tubular, and from 14 to 16 square feet of tube surface is allowed per nominal horse power; but such boilers although they may give abundance of steam are generally needlessly bulky, and the method of fixing the proportions does not appear so eligible as that previously suggested. In sea-going steamers the funnel plates are usually about 9 ft. long and $\frac{3}{16}$ths thick, and where different flues or boilers have their debouch in the same chimney, it is expedient to run divsion plates up the chimney for a considerable distance to keep the draughts distinct. The dampers should not be in the chimney, but at the end of the boiler flue, so that they may be available for use if the funnel by accident be carried away. The waste steam pipe should be of the same height as the funnel, so as to carry the waste steam clear of it, for if the waste steam strikes the funnel it will wear the iron into holes; and the waste steam pipe should be made at the bottom with a faucet joint, to prevent the working of the funnel when the vessel rolls from breaking the pipe at the neck. There should be two hoops round the funnel for the attachment of the funnel shrouds, instead of one, so that the funnel may not be carried overboard if one hoop breaks, or if the funnel breaks at the upper hoop from the corrosive action of the waste steam, as sometimes happens. The deck over the steam chest should be formed of an iron plate supported by angle iron beams, and there should be a high angle iron cooming round the hole in the deck through which the chimney ascends, to prevent any water upon the deck from leaking down upon the boiler. Around the lower part of the funnel there should be a sheet-iron casing to prevent any inconvenient dispersion of heat in that situation; and another short piece of casing, of a somewhat larger diameter and riveted to the chimney, should descend over the first casing, so as to prevent the rain or spray which may beat against the chimney from being poured down within the casing upon the top of the boiler. The pipe for conducting away the waste water from the top of the safety valve should lead overboard, and not into the bilge of the ship, as inconvenience arises from the steam occasionally passing through it, if it has its termination in the engine room. The man-hole and mud-hole doors, unless put on from the outside like a cylinder cover with a great number of bolts, should be put on from the inside with cross bars on the outside; and the bolts should be strong, and have coarse threads and square nuts so that the threads may not be overrun, nor the nuts become round by the unskilful manipulations of the firemen, by whom these doors are removed or replaced. If from any imperfection in the roof of a furnace or flue a patch requires to be put upon it, it will be better to let the patch be applied upon the upper rather than upon the lower surface of the plate; as if applied within the furnace a recess will be formed for the lodgment of deposit, which will prevent the rapid transmission of the heat in that part, and the iron will be very liable to be again burned away. A crack in a plate may be closed by boring holes in the direction of the crack, and inserting rivets with large heads, so as to cover up the imperfection. If the top of the furnace be bent down, from the boiler having been accidentally allowed to get short of water, it may be set up again by a screw-jack—a fire of wood having been previously made beneath the injured plate; but it will in general be nearly as expeditious a course to remove the plate and introduce a new one, and the result will be more satisfactory.

145. *Q.*—Is much inconvenience experienced in marine boilers from saline incrustations upon the flues?

A.—Incrustation in boilers at one time caused much more perplexity than it does at present, as it was supposed that in some seas it was impossible to prevent the boilers of a steamer from becoming salted up: but it has now been satisfactorily ascertained that there is very little difference in the saltness of different seas, and that, however salt the water may be, the boiler will be preserved from any injurious amount of incrustation by blowing off, as it is called, very frequently, or by permitting a considerable portion of the super-salted water to escape at short intervals into the sea. Sea water contains about $\frac{1}{33}$d its weight of salt, and in the open air it boils at the temperature of 213·2°; if the proportion of salt be increased to $\frac{2}{33}$ds of the weight of the water, the boiling point will rise to 214·4°; with $\frac{3}{33}$ds of salt the boiling point will be 215·5°; $\frac{4}{33}$ds, 216·7°; $\frac{5}{33}$ds, 217·9°; $\frac{6}{33}$ds, 219°; $\frac{7}{33}$ds, 220·2°; $\frac{8}{33}$ds, 221·4°; $\frac{9}{33}$ds, 222·5°; $\frac{10}{33}$ds, 223·7°; $\frac{11}{33}$ds, 224·9°; and $\frac{12}{33}$ds, which is the point of saturation, 226°. In a steam boiler the boiling points of water containing these proportions of salt must be higher, as the elevation of temperature due to the pressure of the steam has to be added to that due to the saltness of the water: the temperature of steam at the atmospheric pressure being 212°, its temperature at a pressure of 15 lbs. per square inch will be 250°; and adding to this 4·7° as the increased temperature due to the saltness of the water when it contains $\frac{4}{33}$ds of salt, we have 254·7° as the temperature of the water in the boiler, when it contains $\frac{4}{33}$ds of salt and the pressure of the steam is 15 lbs. on the square inch. It is found by experience that when the concentration of the salt water in a boiler is prevented from exceeding that point at which it contains $\frac{4}{33}$ds its weight of salt, no injurious incrustation will take place; and as sea water contains only $\frac{1}{33}$d of its weight of salt, it is clear that it must be reduced by evaporation to one-fourth of its bulk before it can contain $\frac{4}{33}$ds of salt; or, in other words, a boiler must blow out into the sea one-fourth of the water it receives as feed, in order to prevent the water from rising above $\frac{4}{33}$ds of concentration. Taking the latent heat of steam at 1000° at the temperature of 212°, and reckoning the sum of the latent and sensible heats as forming a constant quantity, the latent heat of steam at the temperature of 250° will be 962°, and the total heat of the steam will be 1212° in the case of fresh water; but as the feed water is sent into the boiler at the temperature of 100°, the accession of heat it receives from the fuel will be 1112° in the case of fresh water, or 1112° increased by 3·98° in the case of containing $\frac{4}{33}$ds of salt—the 3·98° being the 4·7° increase of temperature due to the presence of $\frac{4}{33}$ds of salt, multiplied by 0·847, the specific heat of steam. This makes the total accession of heat received by the steam in the boiler equal to 1115·98°, or say 111 6, which multiplied by 3, as 3 parts of the water are raised into steam, gives us 3348° for the heat in the steam, while the accession of heat received in the boiler by the 1 part of residual brine will be 154·7°, multiplied by 0·85, the specific heat of the brine, or 130·495° ; and 3348° divided by 130·495° is about $\frac{1}{26}$th. It appears, therefore, that by blowing off the boiler to such an extent, that the saltness shall not rise above what answers to $\frac{4}{33}$ds of salt, about $\frac{1}{26}$th of the heat is blown into the sea: this is but a small proportion, and as there will be a greater waste of heat, if from the existence of scale upon the flues the heat can be only imperfectly transmitted to the water, there cannot be even an economy of fuel in niggard blowing off, while it involves the introduction of other evils. To save a part of the heat lost by the operation of blowing off, the hot brine is sometimes passed through a number of small tubes surrounded by the feed water; but there is scarcely any gain from the use of such apparatus, and the tubes are apt to become choked up, whereby the safety of the boiler may be endangered by the injurious concentration of its contents. Pumps, worked by the engine for the extraction of the brine, are generally used in connection with the small tubes for the extraction of the heat from the super-salted water; and if the tubes become choked, the pumps will cease to eject the water, while the engineer may consider them to be all the while in operation. The general mode of blowing off the boiler is to allow the water to rise gradually for an hour or two above the lowest working level, and then to open the cock communicating with the sea, and keep it open until the surface of the water within the boiler has fallen several inches; but in some cases a cock of smaller size is allowed to run water continuously, and in other cases brine pumps are used, as already mentioned: but in every case in which the super-salted water is discharged from the boiler in a continuous stream, a hydrometer or salt gauge of some convenient construction should be applied to the boiler, so that the density of the water may at all times be visible. Blowing off from a point near the surface of the water is more beneficial than blowing off from the bottom of the boiler. Solid particles of any kind, it is well known, if introduced into boiling water, will lower the boiling point in a slight degree, and the steam will chiefly be generated on the surface of the particles, and indeed will have the appearance of coming out of them: if the particles be small, the steam generated beneath and around them will balloon them to the surface of the water, where the steam will be liberated and the particles will descend; and the impalpable particles in a marine boiler, which by their subsidence upon the flues concrete into scale, are carried in the first instance to the surface of the water, so that if they be caught there, and ejected from the boiler, the formation of scale will be prevented. Advantage is taken of this property in Lamb's Scale Preventer, which is substantially a contrivance for blowing off from the surface of the water that in practice is found to be very effectual; but a float in connection with a valve at the mouth of the discharging pipe is there introduced, so as to regulate the quantity of water blown out by the height of the water level, or by the extent of opening given to the feed cock; the operation, however, of the contrivance would be much the same if the float were dispensed with. In some boilers sheet-iron vessels called sediment collectors are employed, which collect into them the impalpable matter which in Lamb's apparatus is ejected from the boiler at once. One of these vessels, of about the size and shape of a loaf of sugar, is put into each boiler, with the apex of the cone turned downwards into a pipe leading overboard, for conducting the sediment away from the boiler. The base of the cone stands some distance above the water line, and in its side conical slits are cut, so as to establish a free communication between the water within the conical vessel and the water outside it. The particles of stony matter, which are ballooned to the surface by the steam in every other part of the boiler, subside within the cone, where no steam is generated, and the water is consequently tranquil; and the deposit is discharged overboard at intervals by means of the cock communicating with the sea. By blowing off from the surface of the water, the requisite cleansing action is obtained with less waste of heat; and where the water is muddy, the foam upon the surface of the water is ejected from the boiler—thereby removing one of the chief causes of priming. As it is very desirable that boilers acting on the principal of continuously blowing off a smal stream of the super-salted water, should be provided with a salt gauge which will give immediate notice of any interruption of the operation, various contrivances have been devised for this purpose, the most of which operate on the principal of a hydrometer; but perhaps a more satisfactory principal would be that of a differential steam gauge, which shall indicate the difference of pressure between the steam in the boiler and the steam of a small quantity of fresh water enclosed in a suitable vessel, and immerged in the water of the boiler. If blowing off be sufficiently practiced, the scale upon the flues will never be much thicker than a sheet of writing paper, and *no excuse* should be accepted from engineers for permitting a boiler to be damaged by the accumulation of calcareous deposit. Flue boilers generally require to be blown off once every watch, or once in the two hours; but tubular boilers may require to be blown off once every twenty minutes, and such an amount of blowing off should in every case be adopted, as will effectually prevent any injurious amount of incrustation. Even with judicious management, however, the boilers may sometimes require to be scaled, and the best method of performing this operation appears to be the following:—Lay a train of shavings along the flues, open the safety valve to prevent the existence of any pressure within the boiler, and light the train of shavings, which, by expanding rapidly the metal of the flues, while the scale from its imperfect conducting power can only expand slowly, will crack off the scale; by washing down the flues with a hose the scale will be carried to the bottom of the boiler, or issue with the water from the mud-hole doors. This method of scaling must be practiced only by the engineer himself, and must not be intrusted to the firemen, who in their ignorance might damage the boiler by overheating the plates. It is only where the incrustation upon the flues is considerable, that this method of removing it need be practiced; in other cases the scale may be chipped off by a hatchet-faced hammer, and the flues may then be washed down with the hose in the manner before described. In tubular boilers a good deal of care is required to prevent the ends of the tubes next the furnace from becoming coated with scale. Even when the boiler is tolerably clean in other places the scale will collect here; and in many cases where the amount of blowing off previously found to suffice for flue boilers has been adopted, an incrustation five-eighths of an inch in thickness has formed in twelve months round the furnace ends of the tubes, and the stony husks enveloping them have actually grown together in some parts so as totally to exclude the water. When a boiler gets into this state the whole of the tubes must be pulled out, which may be done by a Spanish windlass combined with a pair of blocks; and three men when thus provided will be able to draw out from 50 to 70 tubes per day—those tubes with the thickest and firmest incrustations being of course the most difficult to remove. The act of drawing out the tubes removes the incrustation; but the tubes should afterwards be scraped by drawing them backwards and forwards between two old files, fixed in a vice, in the form of the letter V. The ends of the tubes should then be heated and dressed with the hammer, and plunged while at a blood heat into a bed of sawdust to make them cool soft, so that they may be riveted again with facility. A few of the tubes will be so far damaged at the ends by the act of drawing them out, as to be too short for reinsertion: this result might be to a considerable extent obviated, by setting the tube plates at different angles, so that the several horizontal rows of tubes would not be originally of the same length, and the damaged tubes of the long rows would serve to replace the short ones: but the practice would be attended with other inconveniences. Muriatic acid, or muriate of ammonia, commonly called sal-ammoniac, introduced into a boiler, prevents scale to a great extent; but it is liable to corrode the boiler internally, and also to damage the engine, by being carried over with the steam; and the use of such intermixtures does not appear to be necessary, if blowing off from the surface of the water is largely practiced. The soot which collects in the inside of the tubes of tubular boilers is removed by means of a brush, like a large bottle brush; and the carbonaceous scale, which remains adhering to the interior of the tubes, is removed by a circular scraper. Ferules in the tubes interfere with the action of this scraper, and in the case of iron tubes ferules are now generally discarded; but it will sometimes be necessary to use ferules for iron tubes, where the tubes have been drawn and reinserted, as it may be difficult to refix the tubes without such an auxiliary. Tubes one-tenth of an inch in thickness are too thin: one-eighth of an inch is a better thickness, and such tubes will better dispense with the use of ferules, and will not so soon wear into holes

146. *Q.*—Will you explain the nature and cause of priming?

A.—Priming is a violent agitation of the water within the boiler, in consequence of which a large quantity of water passes off with the steam in the shape of froth or spray. Such a result is injurious, both as regards the efficacy of the engine, and the safety of the engine and boiler; for the large volume of hot water carried by the steam into the condenser impairs the vacuum, and throws a great load upon the air pump, which diminishes the speed and available power of the engine; and the existence of water within the cylinder, unless there be safety valves upon the cylinder to permit its escape, will very probably cause some part of the machinery to break, by suddenly arresting the motion of the piston when it meets the surface of the water,—the slide valve being closed to the condenser before the termination of the stroke, in all engines with lap upon the valves, so that the water within the cylinder is prevented from escaping in that direction. At the same time the boiler is emptied of its water too rapidly for the feed pump to be able to maintain the supply, and the flues are in danger of being burnt from a deficiency of water above them. The causes of priming are an insufficient amount of steam room, an inadequate area of water level, an insufficient width between the flues or tubes for the ascent of the steam and the descent of water to supply the vacuity the steam occasions, and the use of dirty water in the boiler. New boilers prime more than old boilers; and steamers entering rivers from the sea are more addicted to priming than if sea or river water had alone been used in the boilers—probably from the boiling point of salt water being higher than that of fresh, whereby the salt water acts like so much molten metal in raising the fresh water into steam. Opening the safety valve suddenly may make a boiler prime, and if the safety valve be situated near the mouth of the steam pipe, the spray or foam thus created may be mingled with the steam passing into the engine, and materially diminish its effective power; but if the safety valve be situated at a distance from the mouth of the steam pipe, the quantity of foam or spray passing into the engine may be diminished by opening the safety valve; and in locomotives, therefore, it is found beneficial to have a safety valve on the barrel of the boiler at a point remote from the steam chest, by partially opening which, any priming in that part of the boiler adjacent to the steam chest is checked, and a purer steam than before passes to the engine. When a boiler primes, the engineer generally closes the throttle valve partially, turns off the injection water, and opens the furnace doors, whereby the generation of steam is checked, and a less violent ebullition in the boiler suffices. Where the priming arises from an insufficient amount of steam room, it may be mitigated by putting a higher pressure upon the boiler, and working more expansively, or by the interposition of a perforated plate between the boiler and the steam chest, which breaks the ascending water and liberates the steam. In some cases, however, it may be necessary to set a second steam chest on the top of the existing one, and it will be preferable to establish a communication with this new chamber by means of a number of small holes, bored through the iron plate of the boiler, rather than by a single large orifice. Where priming arises from the existence of dirty water in the boiler, the evil may be remedied by the use of collecting vessels, or by blowing off largely from the surface; and where it arises from an insufficient area of water level, or an insufficient width between the flues for the free ascent of the steam and the descent of the superincumbent water, the evil may be abated by the addition of circulating pipes in some part of the boiler which will allow the water to descend freely to the place from whence the steam rises, the width of the water spaces being virtually increased by restricting their function to the transmission of a current of steam and water to the surface. It is desirable, however, to arrange the heating surface in such a way that the feed water entering the boiler at its lowest point is heated gradually as it ascends, until towards the superior part of the flues it is raised gradually into steam; and in boilers designed upon this principle, there will be less need for any special provision to enable currents to rise or descend. The steam pipe proceeding to the engine should obviously be attached to the highest point of the steam chest, in boilers of every construction.

147. *Q.*—What is the chief cause of boiler explosions?

A.—The chief cause of boiler explosions is, undoubtedly, too great a pressure of steam, or an insufficient strength of boiler; but many explosions have also arisen from the flues having been suffered to become red hot. If the safety valve of a boiler be accidentally jammed, or if the plates or stays be much worn by corrosion while a high pressure of steam is nevertheless maintained, the boiler necessarily bursts; and if from an insufficiency of water in the boiler, or from any other cause, the flues become highly heated, they may be forced down by the pressure of the steam, and a partial explosion may be the result. The worst explosion is where the shell of the boiler bursts, but the collapse of a furnace or flue is also very disastrous generally to the persons in the engine room, and sometimes the shell bursts and the flues collapse at the same time; for if the flues get red hot, and water be thrown upon them either by the feed pump or otherwise, the generation of steam may be too rapid for the safety valve to permit its escape with sufficient facility, and the shell of the boiler may in consequence be rent asunder. Sometimes the iron of the flues becomes highly heated in consequence of the improper configuration of the parts, which by retaining the steam in contact with the metal, prevents the access of the water: the bottoms of large flues upon which the flame beats down, are very liable to injury from this cause, and the iron of flues thus acted upon may be so softened that the flues will collapse upwards with the pressure of the steam. The flues of boilers may also become red hot in some parts from the attachment of scale, which from its imperfect conducting power will cause the iron to be unduly heated; and if the scale be accidentally detached, a partial explosion may occur in consequence. It is found, however, that a sudden disengagement of steam does not immediately follow the contact of water with the hot metal, for water thrown upon red hot iron is not immediately converted into steam, but assumes the spheroidal form, and rolls about in globules over the surface. These globules, however high the temperature of the metal may be on which they are placed, never rise above the temperature of 205°, and give off but very little steam; but if the temperature of the metal be lowered, the water ceases to retain the spheroidal form, and comes into intimate contact with the metal, whereby a rapid disengagement of steam takes place. If water be poured into a very hot copper flask, the flask may be corked up, as there will be scarce any steam produced so long as the high temperature is maintained; but so soon as the temperature is suffered to fall below 350° or 400°, the spheroidal condition being no longer maintainable, steam is generated with rapidity, and the cork will be projected from the mouth of the flask with great force. One useful precaution against the explosion of boilers from too great an internal pressure, consists in the application of a steam gauge to each boiler, which will make the existence of any undue pressure in any of the boilers immediately visible; and every boiler should have a safety valve of its own, the passage leading to which should have no connection with the passage leading to any of the stop valves used to cut off the connection between the boilers; so that the action of the safety valve may be made independent of the action of the stop valve. In some cases stop valves have jammed, or have been carried from their seats into the mouth of the pipe communicating between them, and the action of the safety valves should be rendered independent of all such accidents. Safety valves, themselves, sometimes stick fast from corrosion, from the spindles becoming bent, from a distortion of the boiler top with a high pressure, in consequence of which the spindles become jammed in the guides, and from various other causes which it would be tedious to enumerate; but the inaction of the safety valve is at once indicated by the steam gauge, and, when discovered, the blow-through valves of the engine and blow-off cocks of the boiler should at once be opened, and the fires raked out. A cone in the ball of the waste steam pipe to send back the water carried upwards by the steam, should never be inserted; as in some cases this cone has become loose, and closed up the mouth of the waste steam pipe, whereby the safety valves being rendered inoperative the boiler was in danger of bursting. If the water be carried out of the boiler so rapidly by priming that the level of the water cannot be maintained, and the flues or furnaces are in danger of becoming red hot, the best plan is to open every furnace door and throw in a few buckets full of water upon the fire, taking care to stand sufficiently to the one side to avoid being scalded by the rush of steam from the furnace. There is no time to begin drawing the fires in such an emergency, and by this treatment the fires, though not altogether extinguished, will be rendered incapable of doing harm. If the flues be already red hot, on no account must cold water be suffered to enter the boiler, but the heat should be maintained in the furnaces, and the blow-off cocks be opened, or the mud-hole doors loosened, so as to let all the water escape; but at the same time the pressure must be kept quite low in the boiler, so that there will be no danger of the hot flues collapsing with the pressure of the steam. Plugs of fusible metal were at one time in much repute as a precaution against explosion, the metal being so compounded that it melted with the heat of high-pressure steam; but the device, though ingenious, has not been found of any utility in practice. The basis of fusible metal is mercury, and it is found that the compound is not homogeneous, and that the mercury is forced by the pressure of the steam out of the interstices of the metal combined with it, leaving a porous metal which is not easily fusible, and which is therefore unable to perform its intended function. In locomotives, however, and also in some other boilers, a lead rivet is inserted with advantage in the crown of the fire box, which is melted out if the water becomes too low, and thus gives notice of the danger. All boilers in actual use should be proved at least once a year by forcing water into them by the hand feed-pump until the safety valve is lifted, which should be loaded with at least twice the working pressure for the occasion. If a boiler will not stand this test it is not safe, and either its strength should be increased or the working pressure should be diminished.

148. Q.—Will you state something of the peculiarities of structure of locomotive boilers ?

A.—Locomotive boilers consist of three portions—the barrel containing the tubes, the fire box, and the smoke box; of which the barrel, smoke box, and external fire box are always of iron, but the internal fire box is generally made of copper, though sometimes also it is made of iron. The tubes are sometimes of iron, but generally of brass fixed in by ferules. The whole of the iron plates of a locomotive boiler which are subjected to the pressure of steam should be Lowmoor or Bowling plates of the best quality; and the copper should be coarse grained, rather than rich or soft, and be perfectly free from irregularities of structure and lamination. The thickness of the plates composing the barrel of the boiler varies generally from five-sixteenths to three-eighths of an inch, and the plates should run in the direction of the circumference, so that the fibres of the iron may be in the direction of the strain. The diameter of the barrel commonly varies from 3 ft. to 3 ft. 6 inches ; the diameter of the rivets should be from eleven-sixteenths to three-fourths of an inch, and the pitch of the rivets or distance between their centres should be from seventeen-eighths to 2 inches. The thickness of the plates composing the external fire box is in general three-eighths of an inch if the fire box is circular, and from three-eighths to one-half inch if the fire box is square; and the thickness of the internal fire box is in most cases seven-sixteenths if copper, and from three-eighths to seven-sixteenths of an inch if of iron. Circular internal fire boxes, if made of iron, should be welded rather than riveted, as the rivet heads are liable to be burnt away by the action of the fire; and when the fire boxes are square each side should consist of a single plate, turned over at the edges with a radius of 3 inches, for the introduction of the rivets. The space between the external and internal fire boxes forms a water space, which must be stayed every $4\frac{1}{2}$ or 5 inches by means of copper or iron stay-bolts, screwed through the outer fire box into the metal of the inner fire box, and securely riveted within it: iron stay-bolts are as durable as copper, and their superior tenacity gives them an advantage. The tube plates are generally made from five-eighths to three-fourths of an inch thick; but seven-eighths of an inch thick appears to be preferable, as when the plate is thick the holes will not be so liable to change their figure during the process of feruling the tubes: the distance between the tubes should never be made less than three-fourths of an inch, and the holes should be slightly tapered, so as to enable the tubes to hold the tube plates together. The ferules are for the most part made of steel at the fire box end, and of wrought-iron at the smoke box end, though ferules of malleable cast-iron have in some cases been used with advantage; malleable cast-iron ferules are almost as easily expanded when hammered cold upon a mandril, as the common wrought-iron ones are at a working heat. Spring-steel, rolled with a feather-edge, to facilitate its conversion into ferules, is supplied by some of the steel makers of Sheffield, and it appears expedient to make use of steel thus prepared when steel ferules are em ployed. The roof of the internal fire box, whether flat as in Stephenson's engines, or dome-shaped as in Bury's, requires to be stiffened with cross stay-bars; but the bars require to be stronger and more numerous when applied to a flat surface. The ends of these stay-bars rest above the vertical sides of the fire box; and to the stay-bars thus extending across the crown, the crown is attached at intervals by means of stay-bolts. There are projecting bosses upon the stay-bars encircling the bolts at every point where a bolt goes through, but in the other parts they are kept clear of the fire box crown, so as to permit the access of water to the iron; and, with the view of facilitating the ascent of the steam, the bottom of each stay-bar should be sharpened away in those parts where it does not touch the boiler. The internal and external fire boxes are joined together at the bottom by a ⋈ shaped iron, and round the fire-door they are connected by means of a copper ring 1¼ in. thick, and 2 in. broad; the inner fire box being dished sufficiently outwards at that point, and the outer fire box sufficiently inwards, to enable a circle of rivets three-fourths of an inch in diameter passing through the copper ring and the two thicknesses of iron to make a water-tight joint. To find the proper length of bar requisite for the formation of a hoop of any given diameter, add the thickness of the bar to the required diameter, and the corresponding circumference in a table of circumferences of circles is the length of the bar. If the iron be bent edgewise, the breadth of the bar must be added to the diameter; for it is the thickness of the bar measured radially that is to be taken into consideration. In the tires of railway wheels, which have a flange on one edge, it is necessary to add not only the thickness of the tire, but also two-thirds of the depth of the flange; generally, however, the tire bars are sent from the forge so curved, that the plain edge of the tire is concave, and the flange edge convex, while the side which is afterwards to be bent into contact with the cylindrical surface of the wheel is a plane. In this case the addition of the diameter of two-thirds of the depth of the flange is unnecessary; for the curving of the flange edge has the effect of increasing the real length of the bar. When the tire is thus curved, it is only necessary to add the thickness of the hoop to the diameter, and then to find the circumference from a table; or the same result will be obtained by multiplying the diameter thus increased by the thickness of the hoop by 3·1416.

149. *Q.*—Are locomotive boilers provided with a steam chest?

A.—The upper portion of the external fire box is usually formed into a steam chest, which is sometimes dome-shaped, sometimes semi-circular, and sometimes of a pyramidical form, and from this steam chest the steam is conducted away by an internal pipe to the cylinders; but, in other cases, an independent steam chest is set upon the barrel of the boiler, consisting of a plate-iron cylinder, 20 inches in diameter, 2 ft. high, and three-eighths of an inch thick, with a dome-shaped top, and with the seam welded and the edge turned over to form a flange of attachment to the boiler. The pyramidical dome, of the form employed in Stephenson's locomotives, presents a considerable extent of flat surface to the pressure of the steam, and this flat surface requires to be very strongly stayed with angle irons and tension rods; whereas the semi-globular dome of the kind employed in Bury's engines requires no staying whatever. The man-hole, or entrance into the boiler, consists of a circular or oval aperture, of about 15 in. diameter, placed in Bury's locomotive at the apex of the dome, and in Stephenson's upon the front of the boiler, a few inches below the level of the rounded part; and the cover of the man-hole in Bury's engine contains the safety valve seats. In whatever situation this man-hole is placed, the surfaces of the ring encircling the hole, and of the internal part of the door or cover, should be accurately fitted together by scraping or grinding, so that they need only the interposition of a little red lead to make them quite tight when screwed together. Lead or canvas joints, if of any considerable thickness, will not long withstand the action of high pressure steam; and the whole of the joints about a locomotive should be such that they require nothing more than a little paint or putty, or a ring of wire-gauze smeared with white or red lead, to make them perfectly tight. There must be a mud-hole opposite the edge of each water space, if the fire box be square, to enable the boiler to be easily cleaned out, and these holes are most conveniently closed by screwed plugs made slightly taper. A cock for emptying the boiler is usually fixed at the bottom of the fire box; and it should be so placed as to be accessible when the engine is at work, in order that the engine driver may blow off some water if necessary; but it must not be in such a position as to send the water blown off among the machinery, as it might carry sand or grit into the bearings, to their manifest injury. To save the steam which is formed when the engine is stationary, a pipe is usually fitted to the boiler, which, on a cock being turned, conducts the steam into the water in the tender, whereby the feed water is heated, and less fuel is subsequently required. This method of disposing of the surplus steam may be adopted when the locomotive is descending inclines, or on any occasion where more steam is produced than the engine can consume. The fire bars in locomotives have always been a source of trouble, as, from the intensity of the heat in the furnace, they become so hot as to throw off a scale, and to bend under the weight of the fuel. The best alleviation of these evils lies in making the bars deep and thin : 4 inches deep by five-eighths of an inch thick on the upper side, and three-eighths of an inch on the under side, are found in practice to be good dimensions. In some locomotives, a frame carrying a number of fire bars is made so that it may be dropped suddenly by loosening a catch ; but it is found that any such mechanism can rarely be long kept in working order, as the molten clinker, by running down between the frame and the boiler, will generally glue the frame into its place : it is therefore found preferable to fix the frame, and to lift up the bars by the dart used by the stoker, when any cause requires the fire to be withdrawn. The furnace bars of locomotives are always made of malleable iron ; and for every species of boiler malleable iron bars are to be preferred to bars of cast-iron, as they are more durable, and may, from their thinness, be set closer together, whereby the small coal or coke is saved that would otherwise fall into the ash-pit. The ash-box of locomotives is made of plate-iron a quarter thick: it should not be less than 10 in. deep, and its bottom should be about 9 in. above the level of the rails. The chimney of a locomotive is made of plate-iron one-eighth of an inch thick: it is usually of the same diameter as the cylinder, and must not stand more than 14 ft. high above the level of the rails.

150. *Q.*—What is the best diameter for the tubes of locomotive boilers?

A.—Bury's locomotive with 14 in. cylinders contains 92 tubes of $2\frac{1}{8}$th in. external diameter, and 10 ft. 6 in. long; whereas Stephenson's locomotive with 15 in. cylinders, contains 150 tubes of $1\frac{5}{8}$ths external diameter, and 13 ft. 6 in. long. In Stephenson's boiler, in order that the part of the tubes next the chimney may be of any avail for the generation of steam, the draught has to be very intense, which in its turn involves a considerable expenditure of power: and it is questionable whether the increased expenditure of power upon the blast, in Stephenson's long tubed locomotives, is compensated by the increased generation of steam consequent upon the extension of the heating surface. When the tubes are small in diameter they are apt to become partially choked with pieces of coke, but an internal diameter of $1\frac{5}{8}$ may be employed without inconvenience, if the draught be of medium intensity. The intensity of the draught may easily be diminished by partially closing the damper in the chimney, and it may be increased by contracting the orifice of the blast. A variable blast pipe, the orifice of which may be enlarged or contracted at pleasure, is now much used. There are various devices for this purpose, but the best appears to be that adopted in Stephenson's engine, where a conical nozzle is moved up or down within a blast pipe, which is made somewhat larger in diameter than the base of the cone, but with a ring projecting internally, against which the base of the cone abuts when the nozzle is pushed up. When the nozzle stands at the top of the pipe the whole of the steam has to pass through it, and the intensity of the blast is increased by the increased velocity thus given to the steam; whereas when the nozzle is moved downward the steam escapes through the annular opening left between the nozzle and the pipe, as well as through the nozzle itself, and the intensity of the blast is diminished by the enlargement of the opening for the escape of the steam thus made available. In most locomotives the velocity of the draught is such that it would require very long tubes to extract the heat from the products of combustion, if the heat were transmitted through the metal of the tubes with only the same facility as through the iron of ordinary flue boilers, and if it were required at the same time that the heat should be as thoroughly extracted. The Nile steamer, with engines of 110 nominal horses power each, and with two boilers having two independent flues in each, of such dimensions as to make each flue equivalent to 55 nominal horses power, works at 62 per cent. above the nominal power, so that the actual evaporative efficacy of each flue would be equivalent to 89 actual horses power, supposing the engines to operate without expansion; but as the mean pressure in the cylinder is somewhat less than the initial pressure, the evaporative efficacy of each flue may be reckoned equivalent to 80 actual horses power. With this evaporative power there is a calorimeter of 990 square inches, or 12·3 square inches per actual horse power; whereas in Stephenson's locomotive with 150 tubes, if the evaporative power be taken at 200 cubic feet of water in the hour, which makes the engine equal to 200 actual horses power, and the internal diameter of the tubes be taken at thirteen eighths of an inch, the calorimeter per actual horse power will only be 1·1136 square inches; or, in other words, the calorimeter in the locomotive boiler will be 11·11 times less than in the flue boiler for the same power, so that the draught in the locomotive must be 11·11 times stronger, and the ratio of the length of the tube to its diameter 11·11 times greater than in the flue boiler, supposing the heat to be transmitted with only the same facility. The flue of the Nile, as stated in the answer to Question 93, would require to be 35½ in. in diameter, if made of the cylindrical form, and 47¾ ft. long: the tubes of a locomotive if 1⅜th in. in diameter would only require to be 22·19 in. long with the same velocity of draught: but as the draught is 11·11 times faster than in a flue boiler, the tubes ought to be 246·558 inches, or about 20½ ft. long according to this proportion. In practice, however, they are one-third less than this, which reduces the heating surface from 9 to 6 square feet per actual horse power, and this length even is found to be inconvenient. It is greatly preferable therefore to increase the calorimeter, and diminish the intensity of the draught.

151. *Q.*—You have mentioned the existence of the steam gauge, the vacuum gauge, the salt gauge, and the

indicator; what other gauges or instruments are there for telling the state, or regulating the power, of an engine?

A.—There is the counter for telling the number of strokes the engine makes, and the dynamometer for ascertaining the tractive power of steam vessels or locomotives; then there are the gauge cocks, and glass tubes, or floats, for telling the height of water in the boiler; and in pumping engines there is the cataract for regulating the speed of the engine. The counter consists of a train of wheel work, so contrived that by every stroke of the engine, an index hand is moved forward a certain space, whereby the number of strokes made by the engine in any given time is accurately recorded. In most cases the motion is communicated by means of a detent, attached to some reciprocating part of the engine, to a ratchet wheel which gives motion to the other wheels in its slow revolution: but it is preferable to derive the motion from some revolving part of the engine by means of an endless screw, as where the ratchet is used the detent will sometimes fail to carry it round the proper quantity. In the counter contrived by Mr. Adie, an endless screw works into the rim of two small wheels situated on the same axis, but one wheel having a tooth more than the other, whereby a differential motion is obtained; and the difference in the velocity of the two wheels, or their motion upon one another, expresses the number of strokes performed. The endless screw is attached to some revolving part of the engine whereby a rotary motion is imparted to it; and the wheels into which the screw works hang down from it like a pendulum, and are kept stationary by the action of gravity. The dynamometer employed for ascertaining the traction upon railways consists of two flat springs joined together at the ends by links, and the amount of separation of the springs at the centre indicates, by means of a suitable hand and dial, the force of traction. In screw vessels the forward thrust of the screw is measured by a dynamometer constructed on the principle of a weighing machine, in which a small spring pressure at the index will balance a very great pressure where the thrust is applied; and in each case the variations of pressure are recorded by a pencil, on a sheet of paper, carried forward by suitable mechanism, whereby the mean thrust is easily ascertained. The tractive force of paddle wheel steamers is ascertained by a dynamometer fixed on shore, to which the floating vessel is attached by a rope. By means of the glass tubes affixed to the fronts of boilers the height of the water within any of the boilers is readily ascertainable, for the water will stand at the same height in the tube as in the boiler, with which there is a communication maintained both at the top and bottom of the tube by suitable stop-cocks. The gauge cocks are cocks penetrating the boiler at different heights, and which when opened tell whether it is water or steam that exists at the level at which they are respectively inserted. The cocks connecting the glass tube with the boiler should always be so constructed that the tube may be blown through with the steam, to clear it of any internal concretion that may impair its transparency; and the construction of the sockets in which the tube is inserted should be such, that, even when there is steam in the boiler, a broken tube may be replaced with facility. It is unsafe to trust to the glass gauges altogether, as a means of ascertaining the water level, as sometimes they become choked, and it is necessary, therefore, to have gauge cocks in addition; but if the boiler be short of steam, and a partial vacuum be produced within it, the glass gauges become of essential service, as the gauge cocks will not operate in such a case, for though opened, instead of steam and water escaping from them, the air will rush into the boiler. It is expedient to carry a pipe from the lower end of the glass tube downward into the water of the boiler, and a pipe from the upper end upward into the steam in the boiler, so as to prevent the water from boiling down through the tube, as it might otherwise do, and prevent the level of the water from being ascertainable. The average level of water in the boiler should be above the centre of the tube, and the lowest of the gauge cocks should always run water, and the highest should always blow steam. The float for telling the height of water in the boiler is employed only in the case of land boilers, and its action is like that of a buoy floating on the surface, which, by means of a light rod passing vertically through the boiler, shows at what height the water stands. The float is usually formed of stone or iron, and is so counterbalanced as to make its operation the same as if it were a buoy of timber; and it is generally put in connection with the feed valve, so that in proportion as the float rises, the supply of feed water is diminished. The feed water in land boilers is admitted from a small open cistern, situated at the top of an upright or stand pipe set upon the boiler, and in which there is a column of water sufficiently high to balance the pressure of the steam. The water is sent up into the small cistern by the force pump, and any that does not gain admission to the boiler through the valve in the cistern, runs to waste by an overflow shoot. The height of the water in the stand pipe rises and falls with the pressure of the steam, and a float is therefore placed within it to operate upon the damper, which, when the level of the water rises, from the pressure of the steam becoming strong, partially closes the damper, and thus moderates the intensity of the fire. The cataract consists of a small pump-plunger and barrel, set in a cistern of water, the barrel being furnished on the one side with a valve opening inwards, through which the water obtains admission to the pump chamber from the cistern, and on the other by a cock through which, if the plunger be forced down, the water must pass out of the pump chamber. The engine in the upward stroke of the piston, which is accomplished by the preponderance of weight at the pump end of the beam, raises up the plunger of the cataract by means of a small rod, the water entering readily through the valve already referred to; and when the engine reaches the top of the stroke, it liberates the rod by which the plunger has been drawn up, and the plunger then descends by gravity, forcing out the water through the cock, the orifice of which has previously been adjusted, and the plunger in its descent opens the injection valve, which causes the engine to make a stroke. If the cock of the cataract be shut, it is clear that the plunger cannot descend at all, and as in that case the injection valve cannot be opened, the engine must stand still; but if the cock be slightly opened the plunger will descend slowly, the injection valve will slowly open, and the engine will make a gradual stroke as it obtains the water necessary for condensation. The extent to which the cock is open, therefore, will regulate the speed with which the engine works, so that by the use of the cataract, the speed of the engine may be varied to suit the variations in the quantity of water required to be lifted from the mine. In some cases an air cylinder, and in other cases an oil cylinder, is employed instead of the apparatus just described; but the principle on which the whole of these contrivances operate is identical, and the only difference is in the detail.

152. *Q.*—Will you explain the course of procedure in the erection of a pumping engine, such as is used in Cornwall?

A.—Having fixed on the proper situation of the pump in the pit, from its centre measure out the distance to the centre of the cylinder, from which set off all the other dimensions of the house, including the thickness of the walls, and dig out the whole of the included ground to the depth of the bottom of the cellar, so that the bottom of the cylinder may stand on a level with the natural ground of the place, or lower if convenient; for the less the height of the house above the ground, the firmer it will be. The foundations of the walls must be laid at least two feet lower than the bottom of the cellar, unless the foundation be firm rock; and care must be taken to leave a small drain into the pit quite through the lowest part of the foundation of the lever wall, to let off any water that may be spilt in the engine house, or may naturally come into the cellar. If the foundation at that depth does not prove good, you must either go down to a better, if in your reach, or make it good by a platform of wood or piles, or both. Within the house, low walls must be built to carry the cylinder beams, so as to leave sufficient room to come at the holding-down bolts, and the ends of these beams must also be lodged in the wall. The lever wall must be built in the firmest manner, and run solid, course by course, with thin lime mortar, care being taken that the lime has not been long slaked. If the house be built of stone, let the stones be large and long, and let many headers be laid through the wall: it should also be a rule, that every stone be laid on the broadest bed it has, and never set on its edge. A course or two above the lintel of the door that leads to the condenser, build in the wall two parallel flat thin bars of iron equally distant from each other, and from the outside and inside of the wall, and reaching the whole breadth of the lever wall. About a foot higher in the wall, lay at every four feet of the breadth of the front, other bars of the same kind at right angles to the former course, and reaching quite through the thickness of the wall; and at each front corner lay a long bar in the middle of the side walls, and reaching quite through the front wall: if these bars are 10 ft. or 12 ft. long it will be sufficient. When the house is built up nearly to the bottom of the opening under the great beam, another double course of bars is to be built in, as has been directed. At the level of the upper cylinder beams, holes must be left in the walls for their ends, with room to move them laterally, so that the cylinder may be got in; and smaller holes must be left quite through the walls for the introduction of iron bars, which being firmly fastened to the cylinder beams at one end, and screwed at the other or outer end, will serve, by their going through both the front and back walls, to bind the house more firmly together. The spring beams or iron bars fastened to them, must reach quite through the back wall, and be keyed or screwed up tight; and they must be firmly fastened to the lever wall on each side, either by iron bars, firm pieces of wood, or long strong stones, reaching far back into the wall. They must also be bedded solidly, and the residue of the opening must be built up in the firmest manner. If there be no water in the neighborhood that can be employed for the purpose of condensation, it will be necessary to make a pond, dug in the earth, for the reception of the water delivered by the air pump, to the end that it may be cooled and used again for the engine. The pond may be 3 ft. or 4 ft. deep, and lined with turf, puddled, or otherwise made water tight. Throwing up the water into the air in the form of a jet to cool it, has been found detrimental; as the water is then charged with air which vitiates the vacuum. To pack the piston, take 60 common-sized white or untarred rope-yarns, and with them plait a gasket or flat rope as close and firm as possible, tapering for 18 in. at each end, and long enough to go round the piston, and overlapped for that length; coil this rope the thin way as hard as possible, and beat it with a sledge hammer until its breadth answers the place ; put it in, and beat it down with a wooden drift and a hand mallet ; pour some melted tallow all around ; then pack in a layer of white oakum half an inch thick, so that the whole packing may have the depth of 5 to 6 inches, depending on the size of the engine ; finally, screw down the junk ring. The packing should be beat solid, but not too hard, otherwise it will create so great a friction as to prevent the easy going of the engine. Abundance of tallow should be allowed, especially at first ; the quantity required will be less as the cylinder grows smooth. In some of the more modern pumping engines, the piston is provided with metallic packing, consisting for the most part of a single ring with a tongue-piece to break the joint, and packed behind with hemp. The upper edge of the metallic ring is sharpened away from the inside so as to permit more conveniently the application of hemp packing behind it ; and the junk ring is made much the same as if no metallic packing were employed. To set the engine going, the steam must be raised until the pressure in the steam pipe is at least equal to three pounds on the square inch ; and when the cylinder jacket is fully warmed, and steam issues freely from the jacket cock, open all the valves or regulators : the steam will then forcibly blow out the air or water contained in the eduction pipe ; and to get rid of the air in the cylinder, shut the steam valve after having blown through the engine for a few minutes. The cold water round the condenser will condense some of the steam contained in the eduction pipe, and its place will be supplied by some of the air from the cylinder. The steam valve must again be opened to blow out that air, and the operation is to be repeated until the air is all drawn out of the cylinder. When that is the case shut all the valves, and observe if the vacuum gauge shows a vacuum in the condenser. When there is a vacuum equivalent to three inches of mercury, open the injection a very little, and shut it again immediately; and if this produces any considerable vacuum, open the exhausting valve a very little way, and the injection at the same time. If the engine does not now commence its motion, it must be blown through again until it moves. If the engine be lightly loaded, or if there be no water in the pumps, the throttle valve must be kept nearly closed, and the top and exhaustion regulators must be opened only a very little way, else the engine will make its stroke with violence, and perhaps do mischief. If there is much unbalanced weight on the pump end, the plug which opens the steam valve must be so regulated, that the valve will only be opened very slightly; and if, after a few strokes, it is found that the engine goes out too slowly, the valve may be then so adjusted as to open wider. The engine should always be made to work full stroke, that is, until the catch pins be made to come within half an inch of the springs at each end, and the piston should stand high enough in the cylinder when the engine is at rest, to spill over into the perpendicular steam pipe any water which may be condensed above it; for if water remain upon the piston, it will increase the consumption of steam. When the engine is to be stopped, shut the injection valve and secure it, and adjust the tappets so as to prevent the exhausting valve from opening, and to allow the steam valve to open and remain open; otherwise a partial vacuum may arise in the cylinder, and it may be filled with water from the injection or from leaks. A single acting engine, when it is in good order, ought to be capable of going as slow as one stroke in ten minutes, and as fast as ten strokes in one minute; and if it does not fulfil these conditions, there is some fault which should be ascertained and remedied. In the modern Cornish engines the steam is used very expansively, and a high pressure of steam is employed. In some cases a double-cylinder engine is used, in which the steam, after having given motion to a small piston on the principle of a high pressure engine, passes into a larger cylinder, where it operates on the principle of a condensing engine; but there is no superior effect gained by the use of two cylinders, and there is greater complexity in the apparatus. Instead of the lever walls, cast-iron columns are now frequently used for supporting the main beam; and the cylinder end of the main beam is generally made longer than the pump end, so as to enable the cylinder to have a long stroke, and the piston to move quickly, without communicating such a velocity to the pump buckets as will make them work with such a shock as to wear themselves out quickly. A high pressure of steam, too, can be employed where the stroke is long, without involving the necessity of making the working parts of such large dimensions as would otherwise be necessary; for the strength of the parts of a single acting engine will require to be much the same, whatever the length of the stroke may be. The pump now universally preferred is the plunger pump, which admits of being packed or tightened while the engine is at work; but the lowest lift of a mine is generally supplied with a pump on the suction principle, both with the view of enabling the lowest pipe to follow the water with facility as the shaft is sunk deeper, and to obviate the inconvenience of the valves of the pump being rendered inaccessible by any flooding in the mine. The pump valves of deep mines are a perpetual source of expense and trouble, as, from the pressure of water upon them, it is difficult to prevent them from closing with violence; and many expedients have been contrived to mitigate the evil, of which the valve known as Harvey and West's valve has perhaps gained the widest acceptation. This valve is a compromise between the equilibrium valve, of the kind employed for admitting the steam to and from the cylinder in single acting engines, and the common spindle valve formerly used for that purpose; and to comprehend its action, it is necessary that the action of the equilibrium valve should first be understood. This valve consists substantially of a cylinder open at both ends, and capable of sliding upon a stationary piston fixed upon a rod the length of the cylinder, which proceeds from the centre of the orifice the valve is intended to close. It is clear, that when the cylinder is pressed down until its edge rests upon the bottom of the box containing it, the orifice of the pipe must be closed, as the steam can neither escape past the edge of the cylinder nor between the cylinder and the piston; and it is equally clear, that as the pressure upon the cylinder is equal all around it, and the whole of the downward pressure is maintained by the stationary piston, the cylinder can be raised or lowered without any further exertion of force than is necessary to overcome the friction of the piston and of the rod by which the cylinder is raised. Instead of the rubbing surface of a piston, however, a conical valve-face between the cylinder and piston is employed, which is tight only when the cylinder is in its lowest position; and there is a similar face between the edge of the cylinder and the bottom of the box in which it is placed. The moving part of the valve, too, instead of being a perfect cylinder, is bulged outwards in the middle, so as to permit the steam to escape past the stationary piston when the cylindrical part of the valve is raised. It is clear, that if such a valve were applied to a pump, no pressure of water within the pump would suffice to open it, neither would any pressure of water above the valve cause it to shut with violence; and if an equilibrium valve, therefore, be used as a pump valve at all, it must be opened and shut by mechanical means. In Harvey and West's valves, however, the equilibrium principle is only partially adopted; the lower face is considerably larger in diameter than the upper face, and the difference constitutes an annulus of pressure, which will cause the valve to open or shut with the same force as a spindle valve of the area of the annulus. To deaden the shock still more effectually, the lower face of the valve is made to strike upon end wood driven into an annular recess in the pump bucket; and valves thus constructed work with very little noise or tremor; but it is found in practice, that the use of Harvey and West's valve, or any contrivance of a similar kind, adds materially to the load upon the pump. In some cases canvas valves similar to those referred to in page 68 are used for pumps with the effect of materially mitigating the shock ; but they require frequent renewal, and are of inferior eligibility in their action to the side valve, which might be applied to pumps without inconvenience. The centrifugal pump, however, threatens to supersede pumps of every other kind ; and if the centrifugal pump be employed, there will be no necessity for pump valves at all. Indeed, it appears probable, that by working a common reciprocating pump at a high speed, a continuous flow of water might be maintained through the pipes in such a way as to render the existence of any valves superfluous. The best form of the centrifugal pump appears to be that in which the arms diverge from the bottom, like the letter V. Such pumps both draw and force : and by arranging them in a succession of lifts in the shaft of the mine, the water may be drawn without inconvenience from any depth. The introduction of the centrifugal pump will obviously extinguish the single acting engine, as rotative engines working at a high speed will be the most appropriate form of engine where the centrifugal pump is employed. The single acting engine, indeed, is a remnant of engineering barbarism which must now be superseded by more compendious contrivances. The Cornish engines, though rudely manufactured, are very expensive in production, as a large engine does but little work; whereas, by employing a smaller engine, moving with a high speed, the dimensions may be so far diminished that the most refined machinery may be obtained at less than the present cost. It is a mistake to suppose that there is any peculiar virtue in the exisiting form of Cornish engine to make it economical in fuel, or that a less lethargic engine would necessarily be less efficient. The large duty of the engines in Cornwall is traceable to the large employment of the principle of expansion, and to a few other causes which may be made of quite as decisive efficacy in smaller engines working with a quicker speed; and there is therefore no argument in the performance of the present engines against the proposed substitution.

153. *Q.*—What description of rotative engine do you consider the best?

A.—The oscillating engine appears to me the best form of rotative engine yet introduced, and I would prefer it to any other in every case where a rotary motion is required. For land purposes it is simpler and more compact than the common beam engine; for steam vessels it appears to be preferable to the side lever engine, as well as to any of the other forms of direct action engines yet introduced; and for locomotives its employment seems to promise the advantage of diminishing the mass of reciprocating material, whereby there will be less of the sinuous or oscillating motion, by which the safety of railway trains is sometimes endangered. The chief varieties of direct action engines used for steam vessels, besides the oscillating engine, are, the Gorgon engine, with the connecting rod reaching from the top of the piston rod to the crank situated above it; the Steeple engine, with the connecting rod above the crank; and the Annular and Siamese engines of Maudslay. The Gorgon engine has the disadvantage of inconveniently raising the shaft to give room for the stroke, whereby a large paddle wheel becomes necessary; and with a large wheel there must either be a great deal of slip, or the engines must work but slowly. The Steeple engine has the inconvenience of protruding a large portion of the machinery above the deck; and the Annular and Siamese engines have peculiar complications, which render them inferior to the oscillating engine for every purpose. Against the oscillating engine itself various objections have been brought at various times:—the cylinder, it was said, would become oval, the trunnion bearings would be liable to heat and the trunnion joints to leak, the strain upon the trunnions would be apt to bend in or bend out the sides of the cylinder, and the circumstance of the cylinder being fixed across its centre, while the shaft requires to accommodate itself to the working of the ship, might be the occasion of such a strain upon the trunnions as would either break them or bend the piston rod. It is a sufficient reply to these objections to say that they are all hypothetical, and that none of them in practice have been found to exist—to such an extent at least as to occasion any inconvenience; but it is not difficult to show that they are altogether unsubstantial, even without a recourse to the disproofs afforded by experience. There is, no doubt, a tendency in oscillating engines for the cylinder and the stuffing-box to become oval, but after a number of years' wear it is found that the amount of ellipticity is less than what is found to exist in the cylinders of side-lever engines after a similar trial. The resistance opposed by friction to the oscillation of the cylinder is so small, that a man is capable of moving a large cylinder with one hand, whereas in the side-lever engine, if the parallel motion be in the least untrue, which is, at some time or other, an almost inevitable condition, the piston is pushed with great force against the side of the cylinder, whereby a large amount of wear and friction is occasioned. The trunnion bearings, instead of being liable to heat like other journals, are kept down to the temperature of the steam by the flow of steam passing through them; and the trunnion packings are not liable to leak when the packings before being introduced are squeezed in a cylindrical mould. In some cases a hollow, or lantern brass, about one-third, or one-fourth, the length of the packing space, and supplied with steam or water by a pipe, is introduced in the middle of the packing, so that if there be any leakage through the trunnion, it will be a leakage of steam or water, which will not vitiate the vacuum: but in ordinary cases this device will not be necessary, and it is not commonly employed. It is clear that there can be no buckling of the sides of the cylinder by the strain upon the trunnions, if the cylinder be made strong enough, and in cylinders of the ordinary thickness such an action has never been experienced; nor is it the fact, that the intermediate shaft of steam vessels, to which part alone the motion is communicated by the engine, requires to adapt itself to the altering forms of the vessel, as the engine and intermediate shaft are rigidly connected, although the paddle shaft requires to be capable of such an adaptation. Even if this objection existed, however, it could easily be met by making the crank pin of the ball and socket fashion, which would permit the position of the intermediate shaft, relatively with that of the cylinder, to be slightly changed, without throwing an undue strain upon any of the working parts.

154. *Q.*—Will you describe the structure of an oscillating engine as made by Messrs. Penn ?

A.—To do this it will be expedient to take an engine of a given power, and then the sizes may be given as well as an account of the configuration of the parts: we may take for an example a pair of engines of 21½ in. diameter of cylinder, and 22 in. stroke, rated by Messrs. Penn at 12 horses power each. The cylinders of this oscillating engine are placed beneath the cranks, and, as in all Messrs. Penn's smaller engines, the piston rod is connected to the crank pin by means of a brass cap, provided with a socket, by means of which it is cuttered to the piston rod. There is but one air pump, which is situated within the condenser between the cylinders, and is wrought by means of a crank in the intermediate shaft—this crank being cut out of a solid piece of metal, as in the formation of the cranked axles of locomotive engines. The steam enters the cylinder through the outer trunnions, or the trunnions adjacent to the ship's sides, and enters the condenser through the two midship trunnions—a short three-ported valve being placed on the front of the cylinder to regulate the flow of steam to and from the cylinder in the proper manner. This valve is balanced by a weight upon the other side of the cylinder, but in the most recent engines this weight is discarded, and two valves are used, which balance one another. The framing consists of an upper and lower frame of cast-iron, bound together by eight malleable iron columns: upon the lower frame the pillow blocks rest which carry the cylinder trunnions, and the condenser and the bottom frame are cast in the same piece. The upper frame supports the paddle shaft pillow blocks; and pieces are bolted on in continuation of the upper frame to carry the paddle wheels, which are overhung from the journal. The web, or base plate, of the lower frame is $\frac{3}{4}$ of an inch thick, and a cooming is carried all round the cylinder, leaving an opening of sufficient size to permit the necessary oscillation. The cross section of the upper frame is that of a hollow beam 6 in. deep, and about $3\frac{1}{2}$ in. wide, with holes at the sides to take out the core; and the thickness of the metal is $\frac{13}{16}$ths of an inch. Both the upper and the lower frame is cast in a single piece, with the exception of the continuations of the upper frame, which support the paddle wheels. An oval ring 3 in. wide is formed in the upper frame, of sufficient size to permit the working of the air pump crank; and from this ring feathers run to the ends of the cross portions of the frame which support the intermediate shaft journals. The columns are $1\frac{1}{2}$ in in diameter; they are provided with collars at the lower ends, which rest upon bosses in the lower frame, and with collars at the upper ends for supporting the upper frame; but the upper collars of two of the corner columns are screwed on, so as to enable the columns to be drawn up when it is required to get the cylinders out. The cross section of the bottom frame is also of the form of a hollow beam, 7 in. deep, except in the region of the condenser, where it is, of course, of a different form; the depth of the boss for the reception of the columns is a little more than 7 in. deep on the lower frame, and a little more than 6 in. deep on the upper frame; and the holes through them are so cored out, that the columns only bear at the upper and lower edges of the hole, instead of all through it—a formation by which the fitting of the columns is facilitated. The condenser, which is cast upon the lower frame, consists of an oval vessel 22½ in. wide, by 2 ft. 4¼ in. long, and 1 ft. 10½ in. deep; it stands 9 in. above the upper face of the bottom frame, the rest projecting beneath it; and it is enlarged at the sides by being carried beneath the trunnions. The air pump, which is set in the centre of the condenser, is 15¼ in. in diameter, and has a stroke of 11 in. The foot valve is situated in the bottom of the air pump, and consists of a disc of brass, in which there is a rectangular flap valve, but rounded on one side to the circle of the pump, opening upwards, and so balanced as to enable the valve to open with facility; and the balance weight, which is formed of brass cast in the same piece as the valve itself, operates as a stop, by coming into contact with the disc which constitutes the bottom of the pump when the valve has opened sufficiently. This disc is bolted to the bottom of the pump by means of an internal flange, and before it can be removed the pump must be lifted out of its place. The air pump barrel is of brass, to which is bolted a cast-iron mouth piece, with a port for carrying the water to the hot well; and within the hot well the delivery valve, which consists of a common flap valve, is situated. The mouth piece and the air pump barrel are made tight to the condenser, and to one another, by means of metallic joints carefully scraped to a true surface, so that a little white or red lead interposed, makes an air-tight joint. The air pump bucket is of brass, and the valve of the bucket is of the common pot-lid, or spindle kind. The injection water enters through a single cock in front of the condenser, the jet striking against the barrel of the air pump: the air pump rod is maintained in its vertical position by means of guides, the lower ends of which are bolted to the mouth of the pump, and the upper to the oval in the top frame, within which the air pump crank works, and the motion is communicated from this crank to the pump rod by means of a short connecting rod. The lower frame is not set immediately below the top frame, but 2½ in. behind it, and the air pump and condenser are 2½ in. nearer one edge of the lower frame than the other. The thickness of the metal of the cylinder is $\frac{9}{16}$ths of an inch; the depth of the belt of the cylinder is 9½ in., and its greatest projection from the cylinder is 2½ in. The distance from the lower edge of the belt to the bottom of the cylinder is 11½ in., and from the upper edge of the belt to the top flange of the cylinder is 9 in. The trunnions are 7¼ in. diameter in the bearings, and 3½ in. in width; and the flanges to which the glands are attached for screwing in the trunnion packings are 1½ in. thick, and have $\frac{7}{8}$ths of an inch projection. The width of the packing space round the trunnions is $\frac{5}{8}$ths of an inch, and the diameter of the pipe passing through the trunnions $4\frac{5}{8}$ths, which leaves $\frac{11}{16}$ths for the thickness of the metal of the bearing. The pipe leading to the condenser, from the cylinder, is made somewhat bell-mouthed where it joins the condenser, and the gland for compressing the packing is made of a larger internal diameter in every part except at the point the pipe emerges from it, where it accurately fits the pipe, so as to enable the gland to squeeze the packing. By this construction the gland may be drawn back without being jammed upon the enlarged part of the pipe, and the enlargement of the pipe towards the condenser prevents the air pump barrel from offering any impediment to the free egress of the steam. The gland is made altogether in four pieces: the ring which presses the packing is made distinct from the flange to which the bolts are attached which force the gland against the packing, and both ring and flange are made in two pieces, to enable them to be got over the pipe. The ring is half-checked in the direction of its depth, and is introduced without any other support to keep the halves together than what is afforded by the interior of the stuffing box; and the flange is half-checked in the direction of its thickness, so that the bolts which press down the ring by passing through this half-checked part also keep the segments of the flange together. The bottom of the trunnion packing space is contracted to the diameter of the eduction pipe, so as to prevent the packing from being squeezed into the jacket; but the eduction pipe does not fit quite tight into this contracted part, but, while in close contact on the lower side, has about $\frac{1}{32}$d of an inch of space between the top of the pipe and the cylinder, so as to permit the trunnions to wear to that extent without throwing a strain upon the pipe. The eduction pipe is attached to the condenser by a flange joint, and the bolt-holes are all made somewhat oblong in the perpendicular direction, so as to permit the pipe to be slightly lowered, should such an operation be rendered necessary by the wear of the trunnion bearings; but in practice the wear of the trunnion bearings is found to be so small, as to be almost inappreciable. It is not expedient to cast the trunnion plummer blocks upon the lower frame, as is sometimes done; for the cylinders, being pressed from the steam trunnions by the steam, and drawn in the direction of the condenser by the vacuum, have a continual tendency to approach one another; and as they wear slightly towards midships, there would be no power of readjustment unless the plummer blocks were movable. The flanges of the trunnions should always fit tight against the plummer block sides, but there should be a little play sideways at the necks of the trunnions, so that the cylinder may be enabled to expand when heated, without throwing an undue strain upon the trunnion supports. Above and below each trunnion a feather or bracket runs from the edge of the belt between 3 in. and 4 in. along the cylinder, for the sake of additional support; and in large engines this feather is continued through the interior of the belt, and cruciform feathers are added for the sake of greater stiffness. The projection of the outer face of the trunnion flange from the side of the cylinder is 6½ in.; the thickness of the flange round the mouth of the cylinder is $\frac{3}{4}$ of an inch, and its projection $1\frac{3}{8}$ in.; the height of the cylinder stuffing-box above the cylinder cover is $4\frac{1}{8}$ in., and its external diameter $4\frac{3}{8}$ in.—the diameter of the piston rod being $2\frac{1}{8}$ in.; and the thickness of the stuffing-box flange is $1\frac{1}{8}$ in. The length of the valve casing is $16\frac{1}{2}$ in., and its projection from the cylinder is $3\frac{1}{2}$ in. at the top, $4\frac{1}{4}$ in. at the centre, and $2\frac{1}{2}$ in. at the bottom, so that the back of the valve casing is not made flat, but is formed in a curve. The width of the valve casing is 9 in., but there is a portion the depth of the belt $1\frac{1}{2}$ in. wider, to permit the steam to enter from the belt into the casing. The valve casing is attached to the cylinder by a metallic joint; the width of the flange of this joint is $1\frac{1}{4}$ in., the thickness of the flange on the casing $\frac{1}{2}$ in., and the thickness of the flange on the cylinder $\frac{5}{8}$ths of an inch. The valve is of the ordinary three-ported description, and both cylinder and valve faces are of cast-iron. The projection from the cylinder of the passage for carrying the steam upwards and downwards, from the valve to the top and bottom of the cylinder, is $2\frac{1}{4}$ in., and its width externally $8\frac{5}{8}$ in. The piston is packed with hemp, but the junk ring is made of malleable iron, as cast-iron junk rings have been found liable to break: there are four plugs screwed into the cylinder cover, which, when removed, permit a box-key to be introduced, to screw down the piston packings. The screws in the junk ring are each provided with a small ratchet, cut in a fixed washer upon the head, to prevent the screw from turning back; and the number of clicks given by these ratchets, in tightening up the bolts, enables the engineer to know when they have all been tightened equally. The cap for attaching the piston rod to the crank pin is formed altogether of brass, which brass serves to form the bearing of the crank pin. The external diameter of the socket by which this cap is attached to the piston rod is $3\frac{5}{16}$ in. The diameter of the crank pin is 3 in., and the length of the crank pin bearing $3\frac{7}{8}$ in. The thickness of the brass around the crank pin bearing is 1 in., and the upper portion of the brass is secured to the lower portion by means of lugs, which are of such a depth that the perpendicular section through the centre of the bearing has a square outline measuring 7 in. in the horizontal direction, $3\frac{7}{8}$ in. from the centre of the pin to the level of the top of the lugs, and $2\frac{1}{2}$ in. from the centre of the pin to the level of the bottom of the lugs. The width of the lugs is 2 in., and the bolts passing through them are $1\frac{1}{4}$ in. in diameter. The bolts are tapped into the lower portion of the cap, and are fitted very accurately by scraping where they pass through the upper portion, so as to act as steady pins in preventing the cover of the crank pin bearing from being worked sideways by the alternate thrust on each side. The distance between the centres of the bolts is 5 in., and in the centre of the cover, where the lugs, continued in the form of a web, meet one another, an oil cup $1\frac{5}{8}$ in. in diameter, $1\frac{1}{8}$ in. high, and provided with an internal pipe, is cast upon the cover, to contain oil for the lubrication of the crank pin bearing. The depth of the cutter for attaching the cap to the piston rod is $1\frac{1}{4}$ in., and its thickness is three-eighths of an inch. A similar cap attaches the air pump crank to the connecting rod by which the air pump rod is moved ; but in this instance the diameter of the bearing is 5 in., and the length of the bearing about 3 in. The thickness of the brass encircling the bearing is three-fourths of an inch upon the edge, and $1\frac{1}{8}$ in. in the centre, the back being slightly rounded ; the width of the lugs is $1\frac{5}{8}$ in., and the depth of the lugs is 2 in. upon the upper brass, and 2 in. upon the lower brass, making a total depth of 4 in. The diameter of the bolts passing through the lugs is 1 in., and the bolts are tapped into the lower brass, and accurately fitted into the upper one, so as to act as steady pins, as in the previous instance. The lower eye of the connecting rod is forked, so as to admit the eye of the air pump rod ; and the pin which connects the two together is prolonged into a cross head, the ends of which move in the guides. The forked end of the connecting rod is fixed upon the cross head by means of a feather, so that the cross head partakes of the motion of the connecting rod, and a cap, similar to that attached to the piston rod, is attached to the air pump rod, for connecting it with the cross head. The diameter of the air pump rod is $1\frac{1}{2}$ in. ; the external diameter of the socket encircling the rod is $2\frac{1}{8}$ in., and the depth of the socket $4\frac{1}{2}$ in. from the centre of the cross head. The depth of the cutter for attaching the socket to the rod is 1 in., and its thickness $\frac{5}{16}$ in. The breadth of the lugs is $1\frac{3}{8}$ in., the depth $1\frac{1}{4}$ in., making a total depth of $2\frac{1}{2}$ in. ; and the diameter of the bolts seven-eighths of an inch. The diameter of the cross head at the centre is 2 in., the thickness of each jaw around the bearing 1 in., and the breadth of each $\frac{9}{16}$ in. The diameter of the intermediate shaft journal is $4\frac{3}{16}$ in., and of the paddle shaft journal $4\frac{3}{8}$ in.; the length of the journal in each case is 5 in. The diameter of the large eye of the crank is 7 in., and the diameter of the hole through it is $4\frac{3}{8}$ in.; the diameter of the small eye of the crank is $5\frac{1}{4}$ in., the diameter of the hole through it being 3 in. The depth of the large eye is $4\frac{1}{4}$ in., and of the small eye $3\frac{3}{4}$ in.; the breadth of the web is 4 in. at the shaft end, and 3 in. at the pin head, and the thickness of the web is $2\frac{5}{8}$ in. The width of the notch forming the crank in the intermediate shaft for working the air pump is $3\frac{1}{2}$ in., and the width of each of the arms of this crank is $3\frac{15}{16}$ in.; both the outer and inner corners of the crank are chamfered away, until the square part of the crank meets the round of the shaft. The method of securing the crank pins into the crank eyes of the intermediate shaft consists in the application of a nut to the end of each pin, where it passes through the eye, the projecting end of the pin being formed with a thread, upon which the nut is screwed. The eccentric strap is half an inch thick, and $1\frac{1}{4}$ in. broad; and the flanges of the eccentric, within which the strap works, are each three-eighths of an inch thick. The eccentric is put on in two halves, joined in the diameter of largest eccentricity by means of a single bolt passing through lugs on the central eye, and the back balance is made in a separate piece five-eighths of an inch thick, and is attached by means of two bolts, which also help to bind the halves of the eccentric together. The eccentric rod is attached to the eccentric hoops by means of two bolts passing through lugs upon the rod, and tapped into a square boss upon the hoop; and pieces of iron, of a greater or less thickness, are interposed between the surfaces in setting the valve, to make the eccentric rod of the right length. The eccentric rod is kept in gear by the pull of a small horizontal rod, attached to a vertical blade-spring, and it is thrown out of gear by means of the ordinary disengaging apparatus, which acts in opposition to the spring, as, in cases where the eccentric rod is not vertical, it acts in opposition to the gravity of the rod. The paddle shaft plummer blocks are altogether of brass, and are formed in much the same manner as the cap of the piston rod, only that the sole is flat, as in ordinary plummer blocks, and is fitted between projecting lugs of the framing, to prevent side motion. In the bearings fitted on this plan, however, the upper brass will generally acquire a good deal of play after some amount of wear. The bolts are worked slack in the holes, though accurately fitted at first; and it appears expedient, therefore, either to make the bolts very large, and the sockets through which they pass very deep, or to let one brass fit into the other. The trunnion plummer blocks are formed in the same manner as the shaft plummer blocks; the nuts are kept from turning back by means of a pinching screw passing through a stationary washer.

155. *Q.*—Will you explain in detail the construction of the valve gearing, or such parts of it as are peculiar to the oscillating engine?

A.—The eccentric rod is attached by a pin, 1 in. in diameter, to an open curved link with a tail projecting upwards, and passing through an eye to guide the link in a vertical motion. The link is formed of iron case-hardened, and is 2¾ in. deep at the middle, and 2⅜ in. deep at the ends, and 1 in. broad. The opening in the link, which extends nearly its entire length, is 1$\frac{5}{16}$ in. broad, and into this opening a brass block 2 in. long is truly fitted, there being a hole through the block ¾ in. diameter, for the reception of the pin of the valve shaft lever. The valve shaft is 1¾ in. diameter at the end next the link or segment, and diminishes regularly to the other end; but it assumes the form of an octagon in its passage round the cylinder, measuring midway 1¼ in. deep, by about ¾ in. thick; and the greatest depth of the finger for moving the valve is about 1 in. The depth of the lever for moving the valve shaft is 2 in. at the broad, and 1¼ in. at the narrow end. The internal breadth of the mortise in which the valve finger moves is 1$\frac{5}{16}$ in., and its external depth is 1¾ in., which leaves three-eighths of an inch as the thickness of metal round the hole; and the breadth, measuring in the direction of the hole, is 1⅛ in. The valve rod is three-fourths of an inch in diameter, and the mortise is connected to the valve rod by a socket 1 in. long, and 1⅛ in. diameter, through which a small cutter passes. A continuation of the rod, eleven-sixteenths of an inch diameter, passes upward from the mortise, and works through an eye, which serves the purpose of a guide. In addition to the guide afforded to the segment by the ascending tail, it is guided at the ends upon the columns of the framing by means of thin semicircular brasses, 4 in. deep, passing round the columns, and attached to the segment by two $\frac{3}{8}$ in. bolts at each end, passing through projecting feathers upon the brasses and segment, three-eighths of an inch in thickness. The curvature of the segment is such as to correspond with the arc swept from the centre of the trunnion with the distance from the centre of the trunnion to the centre of the valve lever pin when the valve is at half stroke as a radius; and the operation of the segment is to prevent the valve from being affected by the oscillation of the cylinder; but the same action would be obtained by the employment of a smaller eccentric with more lead. In some engines the hollow segment is not formed in a single piece, but of two curved blades, with blocks interposed at the ends, which may be filed down a little, to enable the sides of the slot to be brought nearer, as the metal wears away.

156. *Q.*—What kind of paddle wheel is supplied with these oscillating engines?

A.—The wheels are of the feathering kind, 9 ft. 8 in. in diameter, measuring to the edges of the floats; and there are 10 floats upon each wheel, measuring 4 ft. 6 in. long each, and 18½ in. broad. There are two sets of arms to the wheel, which converge to a cast-iron centre, formed like a short pipe with large flanges, to which the arms are affixed. The diameter of the shaft, where the centre is put on, is 4½ in., the external diameter of the pipe is 8 in., and the diameter of the flanges is 20 in., and their thickness 1¼ in. The flanges are 12 in. asunder at the outer edge, and they partake of the converging direction of the arms. The arms are 2¼ in. broad, and half an inch thick;

the heads are made conical, and each is secured into a recess upon the side of the flange by means of three bolts. The ring which connects together the arms, runs round at a distance of 3 ft. 6 in. from the centre, and the projecting ends of the arms are bent backwards the length of the lever which moves the floats, and are made very wide and strong at the point where they cross the ring, to which they are each attached by four rivets. The feathering action of the floats is accomplished by means of a pin fixed to the interior of the paddle box, set 3 in. in advance of the centre of the shaft, and in the same horizontal line. This pin is encircled by a cast-iron collar, to which rods are attached $1\frac{3}{8}$ in. diameter in the centre, proceeding to the levers, 7 in. long, fixed on the back of the floats in the line of the outer arms. One of these rods, however, is formed of nearly the same dimensions as one of the arms of the wheel, and is called the driving arm, as it causes the cast-iron collar to turn round with the revolution of the wheel; and this collar, by means of its attachments to the floats, accomplishes the feathering action. The eccentricity in this wheel is not sufficient to keep the floats nearly in the vertical position; but this is of less consequence, as only one float is wholly immersed. The diameter of the pins upon which the float turns is $1\frac{3}{8}$ in., and between the pins and the paddle ring two stud rods are set between each of the projecting ends of the arms, so as to prevent the two sets of arms from being forced nearer or further apart; and thus prevent the ends of the arms from hindering the action of the floats, by being accidentally jammed upon the sides of the joints.

Stays, crossing one another, proceed from the inner flange of the centre to the outer ring of the wheel, and from the outer flange of the centre to the inner ring of the wheel, with the view of obtaining greater stiffness. The floats are formed of plate iron, and the whole of the joints and joint pins are steeled, or formed of steel.

157. *Q.*—Will you give the dimensions of some other oscillating engines?

A.—In Messrs. Penn's 50-horse power oscillating engine, the diameter of the cylinder is 3 ft. 4 in., and the length of the stroke 3 ft. The thickness of the metal of the cylinder is 1 in., and the thickness of the cylinder bottom is 1¾ in., crossed with feathers, to give it additional stiffness. The diameter of the trunnion bearings is 1 ft. 2 in., and the breadth of the trunnion bearings 5½ in. Messrs. Penn, in their larger engines, generally make the area of the steam trunnion less than that of the eduction trunnion, in the proportion of 32 to 37; and the diameter of the eduction trunnion is regulated by the internal diameter of the eduction pipe, which is about one-fifth of the diameter of the cylinder. But a somewhat larger proportion than this appears to be expedient: Messrs. Rennie make the area of their eduction pipes, in oscillating engines, $\frac{1}{22}$d of the area of the cylinder. In the oscillating engines of the Oberon, by Messrs. Rennie, the cylinder is 61 in. diameter, and 1½ in. thick above and below the belt, but in the wake of the belt it is 1¼ in. thick, which is also the thickness of metal of the belt itself. The internal depth of the belt is 2 ft. 6 in., and its internal breadth is 4 in. The piston rod is 6¾ in. in diameter, and the total depth of the cylinder stuffing box is 2 ft. 4 in., of which 18 in. consists of a brass bush; this depth of bearing being employed to prevent the stuffing box or cylinder from wearing oval. It is expedient, in oscillating engines, to form the piston with a projecting rim round the edge above and below, and a corresponding recess in the cylinder cover and cylinder bottom, whereby the breadth of bearing of the solid part of the metal will be increased, and in many engines this is now done. The diameter of cylinders of the oscillating engines of the steamers Pottinger, Ripon, and Indus, by Miller and Ravenhill, is 76 in., and the length of the stroke is 7 ft. The thickness of the metal of the cylinder is $1\frac{11}{16}$ in.; diameter of the piston rod $8\frac{3}{4}$ in.; total depth of cylinder stuffing box 3 ft.; depth of bush in stuffing box 4 in.; the rest of the depth, with the exception of the space for packing, being occupied with a very deep gland, bushed with brass. The internal diameter of the steam pipe is 13 in.; diameter of steam trunnion journal 25 in.; diameter of eduction trunnion journal 25 in.; thickness of metal of trunnions $2\frac{1}{4}$ in.; length of trunnion bearings 11 in.; projection of cylinder jacket, 8 in.; depth of packing space in trunnions 10 in.; width of packing space in trunnions, or space round the pipes, $1\frac{1}{2}$ in.; diameter of crank pin $10\frac{1}{4}$ in.; length of bearing of crank pin $15\frac{1}{2}$ in. There are six boilers on the tubular plan in each of these vessels; the length of each boiler is 10 ft. 6 in., and the breadth 8 ft., and each boiler contains 62 tubes 3 in. in diameter, and 6 ft. 6 in. long, and two furnaces 6 ft. $4\frac{1}{2}$ in. long, and 3 ft. $1\frac{1}{2}$ in. broad. In all oscillating engines of any considerable size, the cover of the connecting brass, which attaches the crank pin to the connecting rod, is formed of malleable iron; and the socket also, which is cuttered to the end of the piston rod, is of malleable iron, and is formed with a T head, through which bolts pass up through the brass, to keep the cover of the brass in its place. The packing of the trunnions, after being plaited as hard as possible, and cut to the length to form one turn round the pipe, is dipped into boiling tallow, and is then compressed in a mould, consisting of two concentric cylinders, with a gland forced down into the annular space by three to six screws in the case of large diameters, and one central screw in the case of small diameters. Unless the trunnion packings be well compressed, they will be likely to leak air, and it is therefore necessary to pay particular attention to this condition. It is also very important that the trunnions be accurately fitted into their brasses by scraping, so that there may not be the smallest amount of play left upon them; for if any upward motion is permitted, it will be impossible to prevent the trunnion packings from leaking.

158. *Q.*—How do you set out the trunnions of oscillating engines, so that they shall be at right angles with the interior of the cylinder?

A.—Having bored the cylinder, faced the flange, and bored out the hole through which the boring bar passes, put a piece of wood across the mouth of the cylinder, and jam it in, and put a similar piece in the hole through the bottom of the cylinder. Mark the centre of the cylinder upon each of these pieces, and put into the bore of each trunnion an iron plate, with a small indentation in the middle to receive the centre of a lathe, and adjusting screws to bring the centre into any required position. The cylinder must then be set in the lathe, and hung by the centres of the trunnions, and a straight edge must be put across the cylinder mouth and levelled, so as to pass through the line in which the centre of the cylinder lies. Another similar straight edge, and similarly levelled, must be similarly placed across the cylinder bottom, so as to pass through the central line of the cylinder, and the cylinder is then to be turned round in the trunnion centres—the straight edges remaining stationary, which will at once show whether the trunnions are in the same horizontal plane as the centre of the cylinder, and if not, the screws of the plates in the trunnions must be adjusted, until the central point of the cylinder just comes to the straight edge, whichever end of the cylinder is presented. To ascertain whether the trunnions stand in a transverse plane, parallel to the cylinder flange, it is only necessary to measure down from the flange to each trunnion centre, and if both these conditions are satisfied, the position of the centres may be supposed to be right. The trunnion bearings are then turned, and are fitted into blocks of wood, in which they run while the packing space s being turned out. Where many oscillating engines are made, a lathe with four centres is used, which makes the use of straight edges in setting out the trunnions superfluous.

159. *Q.*—Is it a beneficial practice to make cylinders with steam jackets?

A.—In Cornwall, where great attention is paid to

economy of fuel, all the engines are made with steam jackets, and, in some cases, a flue winds spirally round the cylinder, for keeping the steam hot. Mr. Watt in his early practice discarded the steam jacket for a time, but resumed it again, as he found its discontinuance occasioned a perceptible waste of fuel; and in modern engines it has been found that where a jacket is used less coal is consumed than where the use of a jacket is rejected. The cause of this diminished effect is not of very easy perception, for the jacket exposes a larger radiating surface for the escape of the heat than the cylinder; nevertheless, the fact has been established beyond doubt by repeated trials, that engines provided with a jacket are more economical than engines without one. The exterior of the cylinder, or jacket, should be covered with several plies of felt, and then be cased in timber, the boards, which must be very narrow, being first dried in a stove, and then bound round the cylinder with hoops, like the staves of a cask. In many of the Cornish engines the steam is let into casings formed in the cylinder cover and cylinder bottom, for the further economization of the heat, and the cylinder stuffing box is made very deep, and a lantern or hollow brass is introduced into the centre of the packing, into which brass the steam gains admission by a pipe provided for the purpose, so that in the event of the packing becoming leaky, it will be steam that will be leaked into the cylinder instead of air, which, being incondensable, would impair the efficiency of the engine. A lantern brass, of a similar kind, is sometimes introduced into the stuffing-boxes of oscillating engines, but its use there is to receive the lateral pressure of the piston rod, and thus take any strain off the packing.

160. *Q.*—Can you give a rule for determining the thickness of the cylinder and of the trunnion bearings in oscillating engines ?

A.—In low pressure engines, whether oscillating or otherwise, the thickness of metal of the cylinder should be about $\frac{1}{40}$th of the diameter of the cylinder, which, with a pressure of steam of 20 lbs. above the atmosphere, will occasion a strain of only 400 lbs. per square inch of section of the metal: the thickness of the metal of the trunnion bearing should be $\frac{1}{32}$d of the diameter of the cylinder, and the breadth of the bearing should be about half its diameter. In high pressure engines the thickness of the cylinder should be about $\frac{1}{16}$th its diameter, which, with a pressure of steam of 80 lbs. upon the square inch, will occasion a strain of 640 lbs. upon the square inch of section of the metal; and the thickness of the metal of the trunnion bearings of high pressure oscillating engines should be $\frac{1}{13}$th of the diameter of the cylinder. It is very doubtful whether the steam trunnions of a high pressure oscillating engine will continue long tight if the packing consists of hemp; and it appears preferable to introduce a brass ring, to embrace the pipe, cut spirally, with an overlap piece to cover the cut, and packed behind with hemp.

161. *Q.*—Will you explain the various stages of the manipulation connected with the preparation of cylinders ?

A.—In the first place the cylinder has to be cast. The mould into which the metal is poured is built up of bricks and loam, which is clay and sand ground together in a mill, with the addition of a little horse dung to give it a fibrous structure and prevent cracks. The loam board, by which the circle of the cylinder is to be swept, is attached to an upright iron bar, at the distance of the radius of the cylinder; and a cylindrical shell of brick is built up, which is plastered on the inside with loam, and made quite smooth by traversing the perpendicular loam board round it. A core is then formed in a similar manner, but so much smaller as to leave a space between the shell and the core equal to the thickness of the cylinder, and into this space the melted metal is poured. Whatever nozzles or projections are required upon the cylinder must be formed by means of wooden patterns, which are built into the shell, and subsequently withdrawn; but where a number of cylinders of the same kind are required, it is advisable to make these patterns of iron, which will not be liable to warp or twist while the loam is being dried. The general ambition in making cylinders is to make them sound and hard; but it is expedient also to make them tough, so as to approach as nearly as possible to the state of malleable iron. This may be done by mixing in the furnace as many different kinds of iron as possible; and it may be set down as a general rule in iron founding, that the greater the number of the kinds of metal entering into the composition of any casting, the denser and tougher it will be. The constituent atoms of the different kinds of iron appear to be of different sizes, and the mixture of different kinds maintains the toughness, while it adds to the density and cohesive power. Hot blast-iron was at one time generally believed to be weaker than cold blast-iron; but it is now questioned whether it is not the stronger of the two. The cohesive strength of unmixed iron is not in proportion to its specific gravity, and its elasticity and power to resist shocks appears to become greater as the specific gravity becomes less. Nos. 3 and 4 are the strongest irons. In most cases, iron melted in a cupola is not so strong as when remelted in an air furnace, and when run into green sand it is not reckoned so strong as when run into dry sand, or loam. The quality of the fuel, and even the state of the weather, exerts an influence in the quality of the iron: smelting furnaces, on the cold blast principle, have long been known to yield better iron in winter than in summer, probably from the existence of less moisture in the air; and it would probably be found to accomplish an improvement in the quality of the iron if the blast were made to pass through a vessel containing muriate of lime, by which the moisture of the air would be extracted; and the expense of such a preparation would not be considerable, as, by subsequent evaporation, the salt might be used over and over again for the same purpose. Before the iron is cast into the mould, the interior of the mould must be covered with finely powdered charcoal, or blackening, as it is technically termed—and the secret of making finely skinned castings lies in using plenty of blackening. In loam and dry sand castings the charcoal should be mixed with thick clay water, and applied until it is an eighth of an inch thick, or more; the surface should be then very carefully smoothed, or sleeked, and if the metal has been judiciously mixed, and the mould thoroughly dried, the casting is sure to be a fine one. Dry sand and loam castings should be, as much as possible, made in boxes: the moulds may thereby be more rapidly and more effectually dried, and better castings will be got with a less expense. The next stage is the boring, and in boring cylinders of 74 in. diameter, the boring bar must move so as to make one revolution in about $4\frac{1}{2}$ minutes, at which speed the cutters will move at the rate of about 5 ft. per minute. In boring brass the speed must be slower; the common rate at which the tool moves in boring brass air pumps is about 3 ft. per minute. If this speed be materially exceeded the tool will be spoiled, and the pump made taper. The speed proper for boring a cylinder will answer for boring the brass air pump of the same engine. A brass air pump of $36\frac{1}{2}$ in. diameter requires the bar to make one turn in about three minutes, which is also the speed proper for a cylinder 60 in. in diameter. To bore a brass air pump $36\frac{1}{2}$ in. in diameter requires a week, an iron one requires 48 hours, and a copper one 24 hours. In turning a malleable iron shaft $12\frac{3}{4}$ in. in diameter the shaft should make about five turns per minute, which is equivalent to a speed in the tool of about 16 ft. per minute. A boring mill, of which the speed may be varied from one turn in six minutes to twenty-five turns in one minute, will be suitable for all ordinary wants that can occur in practice. Messrs. Penn grind their cylinders after they are bored, by laying them on their side, and rubbing a piece of lead smeared with emery and oil, and with a cross iron handle like that of a rolling stone, backwards and forwards—

the cylinder being gradually turned round so as to subject every part successively to the operation. The lead by which this grinding is accomplished is cast in the cylinder, whereby it is formed of the right curve ; but the part of the cylinder in which it is cast should be previously heated by a hot iron, else the metal may be cracked by the sudden heat. In fixing a cylinder into the boring mill great care must be taken that it is not screwed down unequally ; and indeed it will be impossible to bore a large cylinder in a horizontal mill without being oval, unless the cylinder be carefully gauged when standing on end, and be set up by screws when laid in the mill until it again assumes its original form. A large cylinder will inevitably become oval if laid upon its side ; and if, while under the tension due to its own weight, it be bored round, it will become oval again when set upon end. If the bottom be cast in the cylinder it will be probably found to be round at one end and oval at the other, unless a vertical boring mill be employed, or the precautions here suggested be adopted. Nor is it only in the boring of the cylinder that it is necessary to be careful that there is no change of figure; for it will be impossible to face the valves truly in the case of large cylinders, unless the cylinder be placed on end, or internal props be introduced to prevent the collapse due to the cylinder's weight.

162. *Q.*—Have you any thing further to add upon the subject of cylinders?

A.—Nothing that may not be stated in a few words. Locomotive cylinders are generally made an inch longer than the stroke, or there is half an inch of clearance at each end of the cylinder, to permit the springs of the vehicle to act without causing the piston to strike the top or bottom of the cylinder. The thickness of metal of the cylinder ends is usually about a third more than the thickness of the cylinder itself, and both ends are generally made removable. The operation of priming is very injurious to the cylinders and valves of locomotives, especially if the water be sandy, as the grit carried over by the steam wears the rubbing surfaces rapidly away. The face of the cylinder on which the valve works is raised a little above the metal around it, both to facilitate the operation of facing, and with the view of enabling any foreign substance deposited on the face to be pushed aside by the valve into the less elevated part, where it may lie without occasioning any further disturbance. The valve casing is sometimes cast upon the cylinder, and it is generally covered with a door which may be removed to permit the inspection of the faces. In some valve casings the top as well as the back is removable, which admits of the valve and valve bridle being removed with greater facility. A cock is placed at each end of locomotive cylinders, to allow the water to be discharged which accumulates in the cylinder from priming or condensation; and the four cocks of the two cylinders are usually connected together; so that, by turning a handle, the whole are opened at once. In Stephenson's engines, however, with variable expansion, there is but one cock provided for this purpose, which is on the bottom of the valve chest. In all engines the valve casing, if made in a separate piece from the cylinder, should be attached by means of a metallic joint, as such a barbarism as a rust joint in such situations is no longer permissible. In the case of large engines with valve casings suitable for long slides, an expansion joint in the valve casing should invariably be inserted; otherwise the steam, by gaining admission to the valve casing before it can enter the cylinder, expands the casing while the cylinder remains unaltered in its dimensions, and the joints are damaged, and in some cases the cylinder is cracked by the great strain thus introduced. The chest of the blow-through valve is very commonly cast upon the valve casing; and in engines where the cylinders are stationary, this is the most convenient practice. All engines, where the valve is not of such a construction as to leave the face when a pressure exceeding that of the steam is created in the cylinder by priming or otherwise, should be provided with an escape valve to let out the water, and such valve should be so constructed that the water cannot fly out with violence over the attendants; but it should be conducted away by a suitable pipe, to a place where its discharge can occasion no inconvenience. The stuffing boxes of all engines which cannot be stopped frequently to be repacked, should be made very deep: metallic packing in the stuffing box has been used in some engines, consisting in most instances of one or more rings, cut, sprung, and slipped upon the piston rod before the cross head is put on, and packed with hemp behind. This species of packing answers very well when the parallel motion is true, and the piston rod free from scratches, and it accomplishes a material saving of tallow. In some cases a piece of sheet-brass, packed behind with hemp, has been introduced with good effect, a flange being turned over on the under edge of the brass to prevent it from slipping up or down with the motion of the rod. The sheet brass speedily puts an excellent polish upon the rod, and such a packing is more easily kept, and requires less tallow than where hemp alone is employed. In side lever engines the attachments of the cylinder to the diagonal stay are generally made of too small an area, and the flanges are made too thick. A very thick flange cast on any part of a cylinder endangers the soundness of the cylinder, by inducing an unequal contraction of the metal; and it is a preferable course to make the flange for the attachment of the framing thin, and the surface large—the bolts being turned bolts and nicely fitted. If from malformation in this part the framing works to an inconvenient extent, the best expedient appears to be the introduction of a number of steel tapered bolts, the holes having been previously bored out; and if the flanges be thick enough, square keys may also be introduced, half into one flange and half into the other, so as to receive the strain. If the jaw cracks or breaks away, however, it will be best to apply a malleable iron hoop round the cylinder to take the strain, and this will in all cases be the preferable expedient, where, from any peculiarities of structure, there is a difficulty in introducing bolts and keys.

163. *Q.*—Which is the most eligible species of piston?

A.—For large engines, pistons with a metallic packing, consisting of a single ring, with the ends mortised into one another, and a piece of metal let in flush over the joint and riveted to one end of the ring, appears to be the best species of piston; and if the cylinder be oscillating, it will be expedient to chamfer off the upper edge of the ring on the inner side, and to pack it at the back with hemp. If the cylinder be a stationary one, springs may be substituted for the hemp packing; but in any case it will be expedient to make the vertical joints of the ends of the ring run a little obliquely, so as to prevent the joint forming a ridge in the cylinder. For small pistons two rings may be employed, made somewhat eccentric internally to give a greater thickness of metal in the centre of the ring; these rings must be set one above the other in the cylinder, and the joints, which are oblique, must be set at right angles with one another, so as to obviate any disposition of the rings in their expansion, to wear the cylinder oval. The rings must first be turned a little larger than the diameter of the cylinder, and a piece is then to be cut out, so that when the ends are brought together the ring will just enter within the cylinder. The ring, while retained in a state of compression, is then to be put in the lathe and turned very truly, and finally, it is to be hammered on the inside with the small end of the hammer, to expand the metal, and thus increase the elasticity. The rings are then to be fitted laterally to the piston, and to one another, by scraping—a steady pin being fixed upon the flange of the piston, and fitting into a corresponding hole in the lower ring, to keep the lower ring from turning round; and a similar pin being fixed into the top edge of the lower ring to prevent the upper ring from turning round; but the holes into which these pins fit must be made oblong, to enable the rings to press outward as the rubbing surfaces wear. In most cases it will be expedient to press the packing rings out with springs where they are not packed behind with hemp, and the springs should be made very strong, as the prevailing fault of springs is their weakness. Sometimes short bent springs, set round at regular intervals between the packing rings and body of the piston, are employed, the centre of each spring being secured by a steady pin or bolt screwed into the side of the piston; but it will not signify much what kind of spring is used, provided they have sufficient tension. The piston rod, where it fits into the piston, should have a good deal of taper; for if the taper be too small the rod will be drawn through the hole, and the piston will be split asunder. Small grooves are sometimes turned out of the piston rod above and below the cutter hole, and hemp is introduced in order to make the piston eye tight. Most piston rods are fixed to the piston by means of a gib and cutter, but in some cases the upper portion of the rod within the eye is screwed, and it is fixed into the piston by means of an indented nut. This nut is in some cases hexagonal, and in other cases the exterior forms a portion of a cone which completely fills a corresponding recess in the piston; but nuts made in this way become rusted into their seat after some time, and cannot be started again without much difficulty. Messrs. Miller, Ravenhill, and Co. fix in their piston rods by means of an indented hexagonal nut, which may be started by means of an open box-key. The thread of the screw is made flat upon the one side and much slanted on the other, whereby a greater strength is secured, without creating any disposition to split the nut. In side lever engines it is a judicious practice to add a nut to the top of the piston rod, in addition to the cutter for securing the piston rod to the cross head. In a good example of an engine thus provided, the piston rod is 7 in. in diameter, and the screw 5 in., the part of the rod which fits into the cross head eye is 1 ft. 5½ in. long, and tapers from 6½ in. to 6 13/16 in. diameter. This proportion of taper is a good one: if the taper be less, or if a portion of the piston rod within the cross head eye be left untapered, as is sometimes the case, it is very difficult to detach the parts from one another. When pistons are made of a single ring, or of a succession of single rings, the strength of each ring should be tested previously to its introduction into the piston, by means of a lever loaded by a heavy weight. The old practice was to depend chiefly upon grinding as the means of making the rings tight upon the piston or upon one another; but scraping is now chiefly relied on. Some makers, however, finish their steam surfaces by grinding them with powdered Turkey-stone and oil. A slight grinding, or polishing, with powdered Turkey-stone and oil, appears to be expedient in ordinary cases, and may be conveniently accomplished by setting the piston on a revolving table, and holding the ring stationary by a cross piece of wood while the table turns round. Pieces of wood may be interposed between the ring and the body of the piston, to keep the ring nearly in its right position, but these pieces of wood should be fitted so loosely as to give some side play, else the disposition would arise to wear the flange of the piston into a groove. Messrs. Penn's piston for oscillating engines has a single packing ring, with a tongue piece, or mortise end, made in the manner already prescribed. The ring is packed behind with hemp packing, and the piece of metal which covers the joint is a piece of thick sheet copper, and is indented into the iron of the ring, so as to offer no obstruction to the application of the hemp. The ring is fitted to the piston only on the under edge: the top edge is rounded to a point from the inside, and the junk-ring does not bear upon it, but the junk-ring squeezes down the hemp packing between the packing ring and the body of the piston. The variety of pistons employed in locomotives is very great, and sometimes even the more complicated kinds are found to work very satisfactorily; but, in general, those pistons which consist of a single ring and tongue piece, or of two single rings set one above the other, so as to break joint, are preferable to those which consist of many pieces. In Stephenson's pistons the screws are liable to work slack, and the springs to break. The piston rods of all engines are now either case hardened very deeply, or are made of steel; and in locomotive engines the diameter of the piston rod is about one-seventh of the diameter of the cylinder, and is formed of tilted steel. The cone of the piston rod, by which it is attached to the piston, is turned the reverse way to that which is adopted in common engines, with the view of making the cutter more accessible from the bottom of the cylinder, which is made to come off like a door. The top of the piston rod is secured with a cutter into a socket with jaws, through the holes of which a cross head passes, which is embraced between the jaws by the small end of the connecting rod, while the ends of the cross heads move in guides. Between the piston rod clutch and the guide blocks, the feed pump rod joins the cross head in some engines. The guides are formed of steel plates attached to the framing, between which work the guide blocks, fixed on the ends of the cross head, which have flanges bearing against the inner edges of the guides. Steel or brass guides are better than iron ones. Stephenson and Hawthorn attach their guides at one end to a cross stay, at the other to lugs on the cylinder cover; and they are made stronger in the middle than at the ends. Stout guide rods of steel, encircled by stuffing boxes on the ends of the cross head, would probably be found superior to any other arrangement. The stuffing boxes might contain conical brushes cut spirally, in addition to the packing, and a ring, cut spirally, might be sprung upon the rod, and fixed in advance of the stuffing box, with lateral play to wipe the rod before entering the stuffing box, to prevent it from being scratched by the adhesion of dust.

164. *Q.*—Will you explain the method of fitting together the valve and cylinder faces?

A.—Both faces must first be planed, then filed according to the indications of a metallic straight edge, and subsequently of a thick metallic face plate, and finally scraped very carefully until the face plate bears equally all over the surface. In planing any surface, the catches which retain the surface on the planing machine should be relaxed previously to the last cut, to obviate distortion from springing. To ascertain whether the face plate bears equally, smear it over with a little red ochre and oil, and move the face plate slightly, which will fix the color upon the prominent points. This operation is to be repeated frequently, and as the work advances, the quantity of coloring matter is to be diminished, until finally it is spread over the face plate in a thin film, which only dims the brightness of the plate. The surfaces at this stage must be rubbed firmly together to make the points of contact visible, and the higher points will become slightly clouded, while the other parts are left more or less in shade. If too small a quantity of coloring matter be used at first, it will be difficult to form a just conception of the general state of the surface, as the prominent points will alone be indicated; whereas the use of a large quantity of coloring matter in the latter stages, would destroy the delicacy of the test the face plate affords. The scraping tool should be of the best steel, and should be carefully sharpened at short intervals on a Turkey-stone, so as to maintain a fine edge. A flat file bent, and sharpened at the end, makes an eligible scraper for the first stages; and a three-cornered file, sharpened at all the corners, is the best instrument for finishing the operation. The number of bearing points which it is desirable to establish on the surface of the work, depends on the use to which the surface is to be applied, but whether it is to be finished with great elaboration, or otherwise, the bearing points should be distributed equally over the surface. It has already been stated, that in the preparation of valve faces it is necessary to take into account the strain which the metal suffers from its own weight; and, it may be added, that the change of figure consequent thereupon is not instantaneous, but becomes greater after some continuance of the strain than it was at first; so that in gauging a cylinder to ascertain the difference of diameter when it is placed on its side, it should have lain some days upon its side to ensure the accuracy of the operation. Face plates, or planometers, as they are sometimes termed, are supplied by most of the makers of engineering tools: every factory should be abundantly supplied with them, and also with steel straight edges, and there should be a master face plate, and a master straight edge, for the sole use of testing, from time to time, the accuracy of those in use. In fitting the faces of a D valve, great care must be taken that the valve be not made conical: unless the back be exactly parallel with the face, it will be impossible to keep the packing from being rapidly cut away. When the valve is laid upon the face plate, the back must be made quite fair along the whole length, by draw filing, according to the indications of a straight edge; and the distance from the face to the extreme height of the back, must be made identical at each extremity. Should a hole occur either in the valve, in the cylinder, or any other part where the surface requires to be smooth, it may be plugged up with a piece of cast-iron, as nearly as possible of the same texture. Bore out the faulty part, and afterwards widen the hole with an eccentric drill, so that it will be of the least diameter at the mouth. The hole may go more than half through the iron; fit then a plug of cast-iron roughly by filing, and hammer it into the hole, whereby the plug will become riveted in it, and its surface may then be filed smooth. Square pieces may be let in after the same fashion, the hole being made dove-tailed, and the pieces thus fitted will never come out. Brass faces are put upon valves or cylinders by means of small brass screws, tapped into the iron with conical necks for the retention of the brass; they are screwed by means of a square head, which, when the screw is in its place, is cut off and filed smooth. In some cases the face is made of extra thickness, and a rim not so thick runs round it, forming a step or recess for the reception of brass rivets, the heads of which are clear of the face. Much trouble is experienced with every modification of valve face; but cast-iron working upon cast-iron is, perhaps, the best combination yet introduced. A usual practice is to pin brass faces on the cylinder, allowing the valve to retain its cast-iron face. Some makers employ brass valves, and others pin brass on the valves, leaving the cylinder with a cast-iron face. Speculum metal and steel have been tried for the cylinder faces, but only with moderate success. In some cases the brass gets into ruts; but the most prevalent affection is a degradation of the iron, owing to the action of the steam, and the face assuming a granular appearance, something like loaf sugar. This action shows itself only at particular spots and chiefly about the angles of the port or valve face. At first the action is slow; but when once the steam has worked a passage for itself, the cutting away becomes very rapid, and, in a short time, it will be impossible to prevent the engine from heating when stopped, owing to the leakage of steam through the valve into the condenser. Copper steam pipes seem to have some galvanic action on valve faces, and malleable iron pipes have sometimes been substituted;

but they are speedily worn out by oxidation, and the scales of rust which are carried on by the steam scratch the valves and cylinders, so that the use of copper pipes is the least evil. In some engines the valve rod is fitted with a parallel motion, provision being made for the attachment of its standards on the valve cover; but more frequently guides are employed. In those long D valves in which exhaustion is performed from below, it is expedient to cast two projections on the sole plate, to prevent the valve from falling down inconveniently far when the valve links are taken off. In locomotive engines, the valve universally employed is the three-ported valve as already explained.

165. *Q.*—What is the most beneficial construction of slide valve?

A.—The best construction of slide valve appears to be that adopted by Messrs. Penn for their larger engines, and which consists of a three-ported valve, to the back of which a ring is applied of an area equal to that of the exhaustion port, and which, by bearing steam tight against the back of the casing, so that a vacuum may be maintained within the ring, puts the valve in equilibrium, so that it may be moved with an inconsiderable exercise of force. The back of the valve casing is put on like a door, and its internal surface is made very true by scraping. There is a hole through the valve so as to conduct away any steam which may enter within the ring by leakage, and the ring is kept tight against the back of the casing, by means of a ring situated beneath the bearing ring, provided with four lugs, through which bolts pass tapped into bosses on the back of the valve, and, by unscrewing these bolts, which may be done by means of a box key which passes through holes in the casings closed with screwed plugs, the lower ring is raised upwards, carrying the bearing ring before it. The rings must obviously be fitted over a boss upon the back of the valve; and between the rings, which are of brass, a gasket ring is interposed to compensate by its compressibility for any irregularity of pressure, and each of the bolts is provided with a ratchet collar to prevent it from turning back, so that the engineer, in tightening these bolts, will have no difficulty in tightening them equally, if he counts the number of clicks made by the ratchet. Where this species of valve is used, it is indispensable that large escape valves be applied to the cylinder, as a valve on this construction is unable to leave the face.

166. *Q.*—Will you describe the configuration and mode of attachment of the eccentric by which the valve is moved?

A.—In marine engines, the eccentric is loose upon the shaft, for the purpose of backing, and is furnished with a back balance and catches, so that it may stand either in the position for going ahead, or in that for going astern. The body of the eccentric is of cast-iron, and it is put on the shaft in two pieces. The halves are put together with rebated joints to keep them from separating laterally, and they are prevented from sliding out by round steel pins, each ground into both halves: square keys would probably be preferable to round pins in this arrangement, as the pins tend to wedge the jaws of the eccentric asunder. In some cases the halves of the eccentric are bolted together, by means of flanges, which is, perhaps, the preferable practice. The eccentric hoop in marine and land engines is generally of brass; it is expedient to cast an oil cup on the eccentric hoop, and where practicable, a pan should be placed beneath the eccentric for the reception of the oil droppings. The notch of the eccentric rod for the reception of the pin of the valve shaft is usually steeled, to prevent inconvenient wear; for when the sides of the notch wear, the valve movement is not only disturbed, but it is very difficult to throw the eccentric rod out of gear. It is found to be preferable, however, to fit this notch with a brass bush, for the wear is then less rapid, and it is an easy thing to replace this bush with another when it becomes worn. The eccentric catches of the kind usually employed in marine engines, sometimes break off at the first bolt hole, and it is preferable to have a bolt in advance of the catch face, or to have a hoop encircling the shaft with the catches welded on it, the hoop itself being fixed by bolts or a key. This hoop may either be put on before the cranks in one piece, or afterwards in two pieces. In locomotives the structure and attachments of the eccentric differ somewhat from the foregoing. The body of the eccentric here also is of cast-iron, but the eccentric hoops are generally of wrought-iron, as brass hoops are found liable to break. In inside cylinder engines the eccentrics are set on the axle between the cranks, and they are put on in two pieces held together by bolts; but in straight axle engines the eccentrics are cast in a piece, and are secured on the shaft by means of a key. The eccentric, when in two pieces, is retained at its proper angle on the shaft by a pinching screw, which is provided with a jam nut to prevent it from working loose. A piece is left out of the eccentric in casting it to allow of the screw being inserted, and the void is afterwards filled by inserting a dove-tailed piece of metal. Stephenson and Hawthorn leave holes in their eccentrics on each side of the central arm, and they apply pinching screws in each of these holes. The method of fixing the eccentric to the shaft by a pinching screw is scarcely sufficiently substantial, and cases are perpetually occurring, when this method of attachment is adopted, of eccentrics shifting from their place. In the Rouen engines with straight axles, the four eccentrics are cast in one piece. When the eccentric hoops are formed of malleable iron, one-half of the strap is forged with the rod, the other half being secured to it by bolts, nuts, and jam nuts. Pieces of brass are, in some cases, pinned within the malleable iron hoop, but it appears to be preferable to put brasses within the hoop to encircle the eccentric, as in the case of any other bearing. When brass straps are used, the lugs have generally nuts on both sides, so that the length of the eccentric rod may be adjusted by their means to the proper length; but it is better for the lugs of the hoops to abut against the necks of the screws, and if any adjustment be necessary from the wear of the straps, washers can be interposed. In some engines the adjustment is effected by screwing the valve rod, and the cross head through which it passes has a nut on either side of it, by which its position upon the valve rod is determined. The forks of the eccentric rod are of steel; the length of the eccentric rod is the distance between the centre of the crank axle, and the centre of the valve shaft. The valve lever in locomotives is usually longer than the eccentric lever, to increase the travel of the valve. The pins of the eccentric lever wear quickly; Stephenson puts a ferule of brass on these pins, which being loose, and acting like a roller, facilitates the throwing in and out of gear, and when worn can easily be replaced, so that there need be no material derangement of the motion of the valve from play in this situation. The starting lever travels between two iron segments, and can be fixed in any desired position. This is done by a small catch or bell crank, joined to the bottom of the handle at the end of the lever, and coming up by the side of the handle, but pressed out from it by a spring. The smaller arm of this bell crank is jointed to a bolt, which shoots into notches, made in one of the segments between which the lever moves. By pressing the bell crank against the handle of the lever the bolt is withdrawn, and the lever may be shifted to any other point, when the spring being released, the bolt flies into the nearest notch.

167. *Q.*—Will you explain the operation of expansion valves?

A.—Expansion valves have the effect of closing the steam passage leading to the cylinder before the stroke of the piston is completed, whereby the steam shut within the cylinder is enabled to expand; and the mechanical efficacy of a given bulk of steam is increased by the average pressure of the expanding steam, multiplied by the distance through which it has urged the piston. The structure of expansion valves is very various; some are slide valves, but the expansion valve commonly used in marine engines is of the kind used in the Cornish engines, and known as the equilibrium valve ; and it is usually worked by a cam on the shaft. The expansion cam is put on the shaft in two pieces, which are fastened to each other by means of four bolts passing through lugs, and is fixed to the shaft by keys. A roller at one end of a bell-crank lever, which is connected with the expansion valve, presses against the cam so that the motion of the lever will work the valve. The roller is kept against the cam by a weight on a lever attached to the same shaft. If the cam were concentric with the shaft, the lever which presses upon it would remain stationary, and also the expansion valve ; but by the projection of the cam, the end of the lever receives a reciprocating motion, which is communicated to the valve. The position of this projection determines the point in relation to the stroke at which the valve is opened, and its circumferential length determines the length of the time during which the valve continues open. The time at which the valve should begin to open is the same under all circumstances, but the duration of its opening varies with the amount of expansion desired. In order to obtain this variable extent of expansion, there are several projections made upon the cam, each of which gives a different degree, or *grade*, as it is usually called, of expansion. These grades all begin at the same point on the cam, but are of different lengths, so that they would begin to move the lever at the same time, but would differ in the time of returning it to its original position. The change of expansion is effected by moving the roller on to the desired grade; which is accomplished by slipping the lever carrying the roller endways on the shaft or pin sustaining it. In locomotive engines, where the use of cams is inadmissible, other expedients are employed, of which those contrived by Stephenson and by Cabrey operate on the principle of accomplishing the requisite variations of expansion by altering the throw of the slide valve. Stephenson connects the ends of the forward and backward eccentric rods by a link with a curved slot, in which a pin upon the end of the valve rod works. By moving this link so as to bring the forward eccentric rod in the same line with the valve rod, the valve receives the motion due to that eccentric; whereas, if the backward eccentric rod is brought in a line with the valve rod, the valve gets the motion proper for reversing, and if the link be so placed that the valve rod is midway between the two eccentric rods, the valve will remain nearly stationary. Mr. Cabrey makes his eccentric rod terminate in a pin which works into a straight slotted lever, furnished with jaws similar to the jaws on the eccentric rods of locomotives. By raising the pin of the eccentric rod in this slot, the travel of the valve will be varied, and expansive action will be the result. Both Stephenson's and Cabrey's expansion gear are inferior in efficacy to those which operate by the aid of a separate valve. Gonzenbach's arrangement consists of an additional slide valve and valve casing placed on the back of the ordinary slide valve casing, and through this supplementary valve the steam must first pass. This supplementary valve is worked by a double-ended lever, slotted at one end for the reception of a pin on the valve link, the position of which in the slot determines the throw of the supplementary valve, and the consequent degree of expansion. The other end of the lever is provided with a pin, which gears into a notch on the backward eccentric rod, when that rod is not in gear with the steam valve. By this arrangement, while the steam valve receives as usual the motion of the forward eccentric, the expansion valve receives the motion of the reversing eccentric. In the expansion gear of Meyer, the steam valve is a plate perforated by two ports, which are covered by two blocks working upon the back of the valve, and retained on a spindle which screws into both blocks, but in one with a right-handed screw, and in the other with a left-handed screw, so that by turning the spindle the distance between the blocks may be increased or diminished, and the degree of expansion altered accordingly. The amount of expansion may also be varied by means of a movable plate, which, by closing the port more or less, and wire-drawing the steam, will accomplish the amount of expansion which is desired.

168. *Q.*—What are the details of the air pump?

A.—The air pump bucket and valves are all of brass in all modern engines, and the chamber of the pump is lined with copper, or made wholly of brass, whereby a single boring suffices. When a copper lining is used, the pump is first bored out, and a bent sheet of copper is introduced, which is made accurately to fill the place, by hammering the copper on the inside. Air pump rods of Muntz's metal or copper are much used. Iron rods covered with brass are generally wasted away where the bottom cone fits into the bucket-eye, and if the casing be at all porous the water will insinuate itself between the casing and the rod and eat away the iron. If iron rods covered with brass be used, the brass casing should come some distance into the bucket-eye; the cutter should be of brass, and a brass washer should cover the under side of the eye, so as to defend the end of the rod from the salt water. Rods of Muntz's metal are probably on the whole to be preferred. It is a good practice to put a nut on the top of the rod, to secure it more firmly in the cross head eye, where that plan can be conveniently adopted. The part of the rod which fits into the cross head eye, should have more taper when made of copper or brass, than when made of iron; as if the taper be small, the rod may get staved into the eye, whereby its detachment will be difficult. Metallic packing has in some instances been employed in air pump buckets, but its success has not been such as to lead to its further adoption. A deep solid block of metal however, without any packing, has in a few instances been employed with a satisfactory result. Where ordinary packing is employed, the bucket should always be made with a junk ring, whereby the packing may be easily screwed down at any time with facility. The bucket valve is generally of the spindle or pot-lid kind, but butterfly valves are sometimes used. The foot and delivery valves are for the most part of the flap or hanging kind. These valves all make a considerable noise in working, and are objectionable in many ways. Valves on Belidor's construction, which is in effect that of a throttle valve hung off the centre, were some years ago proposed for the delivery and foot valves; and it appears probable that their operation would be more satisfactory than that of the valves usually employed. Some delivery valve seats are bolted into the mouth of the air pump, whereby access to the pump bucket is rendered difficult. If delivery valve seats be put in the mouth of the air pump at all, the best mode of fixing them appears to be that adopted by Messrs. Maudslay. The top of the pump barrel is made quite fair across, and upon this flat surface a plate containing the delivery valve is set, there being a small ledge all round, to keep it steady. Between the bottom of the stuffing box of the pump cover and the eye of the valve seat a short pipe extends encircling the pump rod, its lower end checked into the eye of the valve seat, and its upper end widening out to form the bottom of the stuffing box of the pump cover. Upon the top of this pipe some screws press, which are accessible from the top of the stuffing box gland, and the packing also aids in keeping down the pipe, the function of which is to retain the valve seat in its place. When the pump bucket has to be examined the valve seat may be slung with the cover, so as to come up with the same purchase. For the bucket valves, Messrs. Maudslay employ two or more concentric ring valves, with a small lift. These valves have given a good deal of trouble in some cases, in consequence of the frequent fracture of the bolts which guide and confine the rings; but this is only a fault of detail, which is easily remedied, and the principle appears to be superior to that of any of the other air pump valves at present in common use.

169. *Q.*—What are the most important details of the construction of paddle wheels?

A.—The structure of the feathering wheel has already been described in connection with a description of the oscillating engine; and it will be expedient now to restrict any account of the details to the common radial paddle, as applied to ocean steamers. The best plan of making the paddle centres is with square eyes, and each centre should be secured in its place by means of eight thick keys. The shaft should be burred up against the heads of these keys with a chisel, so as to prevent the keys from coming back of their own accord. If the keys are wanted to be driven back, this burr must be cut off; and if made thick, and of the right taper, they may then be started without difficulty. The shaft must of course be forged with square projections on it, so as to be suitable for the application of centres with square eyes. Messrs. Maudslay and Co. bore out their paddle centres, and turn a seat for them on the shaft, afterwards fixing them on the shaft with a single key. This plan is objectionable for the two reasons, that it is insecure when new, and when old is irremovable. The general practice among the London engineers is to fix the paddle arms at the centre to a plate by means of bolts, a projection being placed upon the plates on each side of the arm, to prevent lateral motion; but this method is inferior in durability to that adopted in the Clyde, in which each arm is fitted into a socket by means of a cutter,—a small hole being left opposite to the end of each arm, whereby the arm may be forced back by a drift. A preferable way would be to form the paddle centre out of the arms themselves, by widening them at the head until they touch one another, and then applying a boiler plate upon one side, and riveting the arms firmly to it. If this plan be adopted, it will be expedient to swell the tops of the uncovered side at the part nearest the centre of the shaft. In the manufacture of this centre, the heads of the arms would first be forged, then placed on the edges, and fitted together on the plate. The holes would then be bored for the rivets, temporary bolts fitted into them, and the key-seats cut, and the ends of the arms pared in the slotting machine. Finally, the arms would be welded on to the heads, and the various parts of the wheel riveted together. Some engineers join the paddle arms to the outer ring by means of bolts; but those bolts after a time generally become slack sideways, and a constant working of the parts of the wheel goes on in consequence. Sometimes the part of the outer ring opposite the arm is formed into a mortise, and the arms are wedged tight in these holes by wedges driven in on each side; but the plan is an expensive one, and not satisfactory, as the wedges work loose even though riveted over at the point. The best mode of making a secure attachment of the arms to the ring, consists in making the arms with long T heads, and riveting the cross piece to the outer ring with a number of rivets, not of the largest size, which would weaken the outer ring too much. The best way of securing the inner rings to the arms is by means of lugs welded on the arms, and to which the rings are riveted. The paddle floats are usually made either of elm or pine : if of the former, the common thickness for large sea-going vessels, is about 2½ inches; if of the latter, 3 inches. The floats should have plates on both sides, else the paddle arms will be very liable to cut into the wood, and the iron of the arms will be very rapidly wasted. When the floats have been fresh put on they must be screwed up several times before they come to a bearing. If this be not done the bolts will be sure to get slack at sea, and all the floats on the weather side may be washed off. It is a good plan to give the thread of the paddle bolts a nick with a chisel, after the nut has been screwed up, which will prevent the nut from turning back. The floats should not be notched to allow of their projection beyond the outer ring, as if the sides of the notch be in contact with the outer ring, the ring is soon eaten away in that part, and the projecting part of the float, being unsupported, is liable to be broken off. It is usual to put a steel plate at each end of the paddle shafts tightened with a key, to prevent end-play when the vessel rolls, but the arrangement is precarious and insufficient. Messrs. Maudslay make their paddle shaft bearings with very large fillets in the corner, with the view of diminishing the evil; but it would be preferable, we conceive, to make the bearings of the crank shafts spheroidal; and, indeed, it would probably be an improvement, if most of the bearings about the engine were to be made in the same fashion. The loose end of the crank pin should be made not spheroidal, but consisting of a portion of a sphere; and a brass bush might then be fitted into the crank eye, that would completely incase the ball of the pin, and yet permit the outer end of the paddle shaft to fall without straining the pin, the bush being at the same time susceptible of a slight end motion. The paddle shaft, where it passes through the vessel's side, is usually surrounded by a lead stuffing box, which will yield if the end of the shaft falls: this stuffing box prevents leakage into the ship from the paddle wheels; but it is expedient, as a further precaution, to have a small tank on the ship's side immediately beneath the stuffing box, with a pipe leading down to the bilge to catch and conduct away any water that may enter around the shaft. The bearing at the outer end of the paddle shaft is sometimes supplied with tallow, forced into a hole in the plummer block cover, as in the case of water wheels; but for vessels intended to perform long voyages, it is preferable to have a pipe leading down to the oil cup above the journal from the top of the paddle box, through which pipe oil may at any time be supplied. The bolts for holding on the paddle floats are made extra strong, on account of the corrosion to which they are subject; and the nuts should be made large, and should be square, so that they may be effectually tightened up, even though their corners be worn away by corrosion. Paddle floats, when consisting of more than one board, should be bolted together edgeways, by means of bolts running through their whole breadth.

170. *Q.*—Will you describe the structure and

arrangement of the pumps, pipes, and cocks of locomotive and marine engines?

A.—The feed pumps of locomotives are generally made of brass, but the plungers are sometimes made of iron, and are generally attached to the piston cross head, though in Stephenson's engines they are worked by rods attached to eyes on the eccentric hoops. There is a ball valve between the pump and the tender, and two usually in the pipe leading from the pump to the boiler, besides a cock close to the boiler, by which the pump may be shut off from the boiler in case of any accident to the valves. The ball valves are guided by four branches, which rise vertically, and join together at the top in a hemispherical form. The shocks of the ball against this cap have in some cases broken it after one week's work, from the top of the cage having been flat, and the branches not having had their junction at the top properly filleted. These valve guards are attached in different ways to the pipes; when one occurs at the junction of two pieces of pipe it has a flange, which, along with the flanges of the pipes and that of the valve seat, are held together by a union joint. It is sometimes formed with a thread at the under end, and screwed into the pipe. The balls are cast hollow to lessen the shock, as well as to save the metal. In some cases where the feed pump plunger has been attached to the cross head, the piston rod has been bent by the strain; and that must in all cases occur, if the communication between the pump and boiler be closed when the engine is started, and there be no escape valve for the water. Spindle valves have in some cases been used instead of ball valves, but they are more subject to derangement; but piston valves, so contrived as to shut a portion of water in the cage when about to close, might be adopted with a great diminution of the shock. Slide valves might easily be applied, and would probably be found preferable to any of the expedients at present in use. It would be a material improvement if the feed pumps were to be set in the tender, and worked by means of a small engine, such as that now used in steam vessels for feeding the boilers. The present action of the feed pumps of locomotives is precarious, as if the valves leak in the slightest degree the steam or boiling water from the boiler will prevent the pumps from drawing. It appears expedient, therefore, that the pumps should be far from the boiler, and should be set among the feed water, so that they will only have to force. If the pumps were arranged in the manner suggested, the boiler could still be fed regularly, though the locomotive was standing still; but it would be prudent to have one pump still wrought in the usual way by the engine, in case of derangement of the other, or in case the pump in the tender might freeze. The pipes connecting the tender with the pumps should allow access to the valves and free motion to the engine and tender. This end is attained by the use of ball and socket joints; and, to allow some end play, one piece of the pipe slides into the other like a telescope, and is kept tight by means of a stuffing box. Any pipe joint between the engine and tender must be made in this fashion. The feed pipe of most locomotive engines enters the boiler near the bottom, and about the middle of its length. In Stephenson's engine the water is let in at the smoke box end of the boiler, a little below the water level: by this means the heat is more fully extracted from the escaping smoke, but the arrangement is of questionable applicability to engines of which the steam dome and steam pipe are at the smoke box end, as in that case the entering cold water would condense the steam. In steam vessels, the feed pump plunger is generally of brass, and the barrel of the pump is sometimes of brass, but generally of cast-iron. There should be a considerable clearance between the bottom of the plunger and the bottom of the barrel, as otherwise the bottom of the barrel may be knocked out, should coal dust or any other foreign substance gain admission, as it probably would do if the injection water were drawn at any time from the bilge of the vessel, as is usually done if the vessel spring a leak. The valves of the feed pump in marine engines are generally of the spindle kind, and are most conveniently arranged in a chest, which may be attached in any accessible position to the side of the hot well. There are two side nozzles upon this chest, of which the lower one leads to the pump, and the upper one to the boiler. The pipe leading to the pump is a suction pipe when the plunger ascends, and a forcing pipe when the plunger descends. The plunger in ascending draws the water out of the hot well through the lowest of the valves, and, in descending, forces it through the centre valve into the space above it, which communicates with the feed pipe. Should the feed cock be shut so as to prevent any feed water from passing through it, the water will raise the topmost valve, which is loaded to a pressure considerably above the pressure of the steam, and escape into the hot well. This arrangement is neater and less expensive than that of having a separate loaded valve on the steam pipe, with an overflow through the ship's side, as is the more usual practice. To enable the boilers to be fed in steam vessels, there is a separate pump provided distinct from the engines, and which may be worked by men standing on the deck by means of appropriate handles; and this pump, in addition to its function of replenishing the boilers, is used to wash the decks, or as a fire engine to quench accidental fire. For that purpose, a double acting feed pump of the plunger kind is preferable to one which operates by a piston. The air vessel of the pump should be furnished with an escape valve to prevent the pump from being split, should it be put in connection with the engine when the cocks in the pipe leading to the boiler are closed, an accident which not unfrequently happens. In this species of pump, the application of a four-way cock enables the pump to draw from the sea, from the boiler, or from the bilge, and the pump can deliver either into the boiler or upon deck. In most of the new vessels fitted with tubular boilers, small engines have been introduced to pump water into the boiler when the vessel stops under steam. Most of these engines are furnished with a crank and fly wheel; but as there is no strain on the fly wheel shaft, the crank is connected to the piston rod by means of a horizontal slot. In some steam vessels floats have been introduced to regulate the feed, but their action cannot be depended on in agitated water, if applied after the common fashion. Floats would probably answer if placed in a cylinder which communicates with the water in the boiler by means of small holes; and a disc of metal might be attached to the end of a rod extending beneath the water level, so as to resist irregular movements from the motion of the ship, which would otherwise impair the action of the apparatus. The admission of feed water into the boiler is sometimes regulated by cocks, and sometimes by spindle valves raised and lowered by a screw. Cocks appear to us to be the preferable expedient, as they are less liable to accident or derangement than screw valves, and in modern steam vessels they are generally employed. The feed water is usually conducted from the feed cock to a point near the bottom of the boiler by means of an internal pipe, the object of this arrangement being to prevent the rising steam from being condensed by the entering water. By being introduced near the bottom of the boiler, the water comes into contact in the first place with the bottoms of the furnaces and flues, and extracts heat from them which could not be extracted by water of a higher temperature, whereby a saving of fuel is accomplished. In some cases the feed water is introduced into a casing around the chimney, from whence it descends into the boiler. This plan appears to be an expedient one when the boiler is short of heating surface, and more than a usual quantity of heat ascends the chimney ; but in well-proportioned boilers a water casing round the chimney is superfluous. When a water casing is used the boiler is usually fed by a head of water, the feed water being forced up into a small tank, from whence it descends into the boiler by the force of gravity, while the surplus runs to waste, as in the feeding apparatus of land engines. The blow-off cocks of a boiler are generally placed some distance from the boiler, but it appears preferable that they should be placed quite close to it, as there are no means of shutting off the water from the pipe between the blow-off cock and the boiler, should fracture or leakage there arise. Every boiler must be furnished with a blow-off cock of its own, independently of the main blow-off cocks on the ship's side, so that the boilers may be blown off separately, and may be shut off from one another. The preferable arrangement appears to be, to cast upon each blow-off cock a bend for attaching the cock to the bottom of the boiler, and the plug should stand about an inch in advance of the front of the boiler, so that it may be removed, or re-ground, with facility. The general arrangement of the blow-off pipes is to put a main blow-off pipe beneath the floor plates, across the ship, at the end of the engines, and into this pipe to lead a separate pipe, furnished with a cock, from each boiler. The main blow-off pipe, where it penetrates the ship's sides, is furnished with a cock: and in modern steam vessels Kingston's valves are also used, which consist of a spindle or plate valve, fitted to the exterior of the ship, so that, if the internal pipe or cock breaks, the external valve will still be operative. Some expedient of this kind is almost necessary, as the blow-off cocks require occasional re-grinding, and the sea cocks cannot be re-ground without putting the vessel into dock, except by the use of Kingston's valves, or some equivalent expedient. All the cocks about an engine should be provided with bottoms and stuffing boxes, and reliance should never be placed upon a single bolt passing through a bottom washer for keeping the plug in its place, in the case of any cock communicating with the boiler; for a great strain is thrown upon that bolt if the pressure of the steam be high, and if the plug be made with much taper, and should the bolt break, or the threads strip, the plug will fly out, and persons standing near may be scalded to death. In large cocks, it appears the preferable plan to cast the bottoms in; and the metal of which all the cocks about a marine engine are made, should be of the same quality as that used in the composition of the brasses, and should be without lead, or other deteriorating material. In some cases the bottoms of cocks are burnt in with hard solder, but this method cannot be depended upon, as the solder is softened and wasted away by the hot salt water, and in time the bottom leaks, or is forced out. The stuffing box of cocks should be made of adequate depth, and the gland should be secured by means of four strong copper bolts. The taper of blow-off cocks is an important element in their construction; as, if the taper be too great, the plugs will have a continual tendency to rise, which, if the packing be slack, will enable grit to get between the faces, while, if the taper be too little, the plug will be liable to jam, and a few times grinding will sink it so far through the shell that the water-ways will no longer correspond. One-eighth of an inch deviation from the perpendicular for every inch in height, is a common angle for the side of the cock, which corresponds with one quarter of an inch difference of diameter in an inch of height; but perhaps a somewhat greater taper than this, or one-third of an inch difference in diameter for every inch of height, is a preferable proportion. The bottom of the plug must be always kept a small distance above the bottom of the shell, and an adequate surface must be left above and below the water-way to prevent leakage. Cocks formed according to these directions will be found to operate satisfactorily in practice, while they will occasion perpetual trouble if there be any malformation. Gauge cocks are generally very inartificially made, and occasion needless annoyance. They are rarely made with bottoms, or with stuffing boxes, and are consequently, for the most part, adorned with stalactites of salt after a short period of service. The water discharged from them, too, from the want of a proper conduit, disfigures the front of the boiler, and adds to the corrosion in the ash pits. It would be preferable to combine the gauge cocks appertaining to each boiler into a single upright tube, connected suitably with the boiler, and the water flowing from them could be directed downwards into a funnel tube communicating with the bilge. The cocks of the glass tubes, as well as of the gauge cocks, should be furnished with stuffing boxes and with bottoms, unless the water enters through the bottom of the plug, which in gauge cocks is sometimes the case. The glass gauge tubes should always be fitted with a cock at each neck communicating with the boiler, so that the water and steam may be shut off if the tube breaks; and the cocks should be so made as to admit of the tubes being blown through with steam to clear them, as in muddy water they will become so soiled that the water cannot be seen. The gauge cocks frequently have pipes running up within the boiler, to the end that a high water-level may be made consistent with an easily accessible position of the gauge cocks themselves. With the glass tubes, however, this species of arrangement is not possible, and the glass tubes must always be placed in the position of the water level. The sea injection cocks are usually made in the same fashion as the sea blow-off cocks, and of about the same size. The injection water is generally admitted to the condenser by means of a slide valve; but a cock appears to be preferable, as it is more easily opened, and has not any disposition to shut of its own accord. The sea injection pipes should be put through the ship's sides in advance of the paddles, so that the water drawn in may not be injuriously charged with air. The waste water pipe passing from the hot well through the vessel's side is provided with a stop valve, called the discharge valve, which is usually made of the spindle kind, so as to open when the water coming from the air pump presses against it. In some cases this valve is a sluice valve, but the hot well is then almost sure to be split, if the engine be set on without the valve having been opened. The opening of the waste water pipe should always be above the load water line, as it will otherwise be difficult to prevent leakage through the engine into the ship when the vessel is lying in harbor. Where the pipes pierce the ship's side, they should be made tight, as follows:— The hole being cut, a short piece of lead pipe, with a broad flange at one end, should be fitted into it, the place having been previously smeared with white lead, and the pipe should then be beaten on the inside, until it comes into close contact all around with the wood. A loose flange should next be slipped over the projecting end of the lead pipe, to which it should be soldered, and the flanges should both be nailed to the timber with scupper nails, white lead having been previously spread underneath. This method of procedure, it is clear, prevents the possibility of leakage down through the timbers; and all, therefore, that has to be guarded against after this precaution, is to prevent leakage into the ship. To accomplish this object, let the pipe which it is desired to attach be put through the leaden hause, and let the space between the pipe and the lead be packed with gasket and white lead, to which a little olive oil has been added. The pipe must have a flange upon it to close the hole in the ship's side; the packing must then be driven in from the outside, and be kept in by means of a gland secured with bolts passing through the ship's side. If the pipe is below the water line, the gland must be of brass; but for the waste water pipe, a cast-iron gland will answer. This method of securing pipes penetrating the side, however, though the best for wooden vessels, will, it is clear, fail to apply to iron ones. In the case of iron vessels, it appears to be the best practice to attach a short iron nozzle, projecting inwards, to the skin, for the attachment of every pipe below the water line, as the copper or brass would waste the iron of the skin if the attachment were made in the usual way. Marine boilers are now generally supplied with stop valves, whereby one boiler may be thrown out of use without impairing the efficacy of the remainder. These stop valves are usually spindle valves of large size, and they are for the most part set in a pipe which runs across the steam chests, connecting the several boilers together. The spindles of these valves should project through stuffing boxes in the covers of the valve chests, and they should be balanced by a weighted lever, and kept in continual action by the steam. If the valves be lifted up, and be suffered to remain up, as is the usual practice, they will become fixed by corrosion in that position, and it will be impossible after some time to shut them on an emergency. These valves should always be easily accessible from the engine room; and it ought not to be necessary for the coal boxes to be empty to gain access to them. The pipes of marine engines should always be made of copper. Cast-iron blow-off pipes have in some cases been employed, but they are liable to fracture, and are dangerous. Every pipe passing through the ship's side, and every pipe fixed at both ends, and liable to be heated and cooled, should be furnished with a faucet or expansive joint; and in the case of cast-iron pipes, the part of the pipe fitting into the faucet should be turned. In the distribution of the faucets of the pipes exposed to pressure, care must be taken that they be so placed, that the parts of the pipe cannot be forced asunder by the strain, as serious accidents have occurred from the neglect of this precaution. In locomotives, the admission of the steam from the boiler to the cylinders is regulated by a valve called the regulator, which is generally placed immediately above the internal fire box, and is connected with two copper pipes; one conducting steam from the highest point of the dome down to it; and the other conducting the steam that has passed through it along the boiler to the upper part of the smoke box. Regulators may be divided into two sorts, viz. those with sliding valves and steam ports, and those with conical valves and seats, of which the latter kind are the best. The former kind have for the most part consisted of a circular valve and face, with radial apertures, the valve resembling the outstretched wings of a butterfly, and being made to revolve on its central pivot by connecting links between its outer edges, or by its central spindle. In some of Stephenson's engines with variable expansion gear, the regulator consists of a slide valve covering a port on the top of the valve chests. A rod passes from this valve through the smoke box below the boiler, and by means of a lever parallel to the starting lever, is brought up to the engineer's reach. Cocks were at first used as regulators, but were given up, as they were found liable to stick fast. A gridiron slide valve has been used by Stephenson, which consists of a perforated square moving upon a face with an equal number of holes. This plan of a valve gives, with a small movement, a large area of opening. In Bury's engines a sort of conical plug is used, which is withdrawn by turning the handle in front of the fire box: a spiral groove of very large pitch is made in the valve spindle, in which fits a pin fixed to the boiler, and by turning the spindle an end motion is given to it, which either shuts or opens the steam passage according to the direction in which it is turned. The best regulator would probably be a valve of the equilibrium description, such as is used in the Cornish engine: there would be no friction in such a regulator, and it could be opened or shut with a small amount of force.

171. *Q.*—Will you enumerate the most interesting details which occur to you in connection with the structure of locomotives?

A.—All locomotives are now made with the framing which supports the machinery situated within the wheels; but for some years a vehement controversy was maintained respecting the relative merits of outside and inside framing, which has terminated however in the universal adoption of the inside framing. A similar diversity of opinion obtains at present as to the relative merits of outside and inside cylinders, the outside cylinders being so designated when placed upon the outside of the framing, with their connecting rods operating upon pins in the driving wheels; while the inside cylinders are situated within the framing, and the connecting rods attach themselves to cranks in the driving axle. The chief objection to outside cylinders is, that they occasion a sinuous motion in the engine which is apt to send the train off the rails; but this action may be made less perceptible or be remedied altogether, by setting the crank pins nearly in the same line instead of at right angles, or by placing a weight upon one side of the wheels, the momentum of which will just balance the momentum of the piston and its connections. The sinuous or rocking motion of locomotives is traceable to the arrested momentum of the piston and its attachments at every stroke of the engine, and the effect of the pressure thus created will be more operative in inducing oscillation the farther it is exerted from the central line of the engine. If both cylinders were set at right angles in the centre of the carriage, and the pistons were both attached to a central crank, there would be no oscillation produced; or the same effect would be realized by placing one cylinder in the centre of the carriage, and two at the sides—the pistons of the side cylinders moving simultaneously; but it is impossible to couple the piston of an upright cylinder direct to the axle of a locomotive, without causing the springs to work up and down with every stroke of the engine: and the use of three cylinders, though adopted in some of Stephenson's later engines, involves too much complication to be a beneficial innovation. It is difficult, in engines intended for the narrow gauge, to get cylinders within the framing of sufficient diameter to meet the exigencies of railway locomotion; by casting both cylinders in a piece, however, a considerable amount of room may be made available to increase their diameters. It is very desirable that the cylinders of locomotives should be as large as possible, so that expansion may be adopted to a large extent; and with any given speed of piston, the power of an engine either to draw heavy loads, or achieve high velocities, will be increased with every increase of the dimensions of the cylinder. The framing of locomotives, to which the boiler and machinery are attached, and which rests upon the springs situated above the axles, is formed generally of malleable iron, but in some engines the side frames consist of oak with iron plates riveted on each side. The guard plates are in these cases generally of equal length, the frames being curved upwards to pass over the driving axle. Hard cast-iron blocks are riveted between the guard plates to serve as guides for the axle bushes. The side frames are connected across at the ends, and cross stays are introduced beneath the boiler to stiffen the frame sideways, and prevent the ends of the connecting or eccentric rods from falling down if they should be broken. The springs are of the ordinary carriage kind, with plates connected at the centre, and allowed to slide on each other at their ends. The upper plate terminates in two eyes, through each of which passes a pin, which also passes through the jaws of the bridle, connected by a double-threaded screw to another bridle, which is jointed to the framing; the centre of the spring rests upon the axle box. Sometimes the springs are placed between the guard plates, and below the framing which rests upon their extremities. One species of spring, which has gained a considerable introduction, consists of a number of flat steel plates, with a piece of metal or other substance interposed between them at the centre, leaving the ends standing apart. It would be preferable, perhaps, to make the plates of a common spring with different curves, so that the leaves, though in contact at the centre, would not be in contact at the ends with light loads, but would be brought into contact gradually, as the strain comes on: a spring would thus be obtained that was suitable for all loads. Behind the locomotive runs another carriage, called a tender, for holding coke and water. A common mode of connecting the engine and tender is by means of a rigid bar, with an eye at each end through which pins are passed. Between the engine and tender, however, buffers should always be interposed, as their pressure contributes greatly to prevent oscillation and other irregular motions of the engine. In most engines a bar is strongly attached to the front of the carriage on each side, and projects perpendicularly downwards to within a short distance of the rail, to clear away stones or other obstructions that might occasion accidents if the engine ran over them. The axles bear only against the top of the axle boxes, which are generally of brass, but a plate extends underneath the bearing, to prevent sand from being thrown up on it. The upper part of the box in most engines has a reservoir of oil, which is supplied to the journal by tubes with siphon wicks. Stephenson uses cast-iron axle boxes with brasses, and grease instead of oil; and the grease is fed upon the journal by the heat of the bearing melting it, whereby it is made to flow down through a hole in the brass. Any engines constructed with outside bearings have inside bearings also, which are supported by longitudinal bars, which serve also in some cases to support the piston guides; these bearings are sometimes made so as not to touch the shafts unless they break. The wheels of a locomotive are always made of malleable iron. The driving wheels are made large to increase the speed; the bearing wheels also are easier on the road when large. In the goods engines the driving wheels are smaller than in the passenger engines, and are generally coupled together. Wheels are made with much variety in their constructive details; sometimes they are made with cast-iron naves, with the spokes and rim of wrought-iron;

but in the best modern wheels, the nave is formed of the ends of the spokes welded together at the centre. When cast-iron naves are adopted, the spokes are forged out of flat bars with T formed heads, and are arranged radially in the founders' mould, the cast-iron, when fluid, being poured among them. The ends of the T heads are then welded together to constitute the periphery of the wheel or inner tire; and little wedge-form pieces are inserted where there is any deficiency of iron. In some cases the arms are hollow, though of wrought-iron; the tire of wrought-iron, and the nave of cast-iron; and the spokes are turned where they are fitted into the nave, and are secured in their sockets by means of cutters. Hawthorn makes his wheels with cast-iron naves and wrought-iron rims and arms; but instead of welding the arms together, he makes palms on their outer end, which are attached by rivets to the rim. These rivets, however, unless very carefully formed, are apt to work loose; and it would probably be found an improvement if the palms were to be slightly indented into the rim, in cases in which the palms do not meet each other at the ends. When the rim is turned it is ready for the tire, which is now made of steel. The materials for wheel tires are first swaged separately, and then welded together under the heavy hammer at the steel works; after which they are bent to the circle, welded, and turned to certain gauges. The tire is now heated to redness in a circular furnace; during the time it is getting hot, the iron wheel, turned to the right diameter, is bolted down upon a face-plate or surface; the tire expands with the heat, and when at a cherry red, it is dropped over the wheel, for which it was previously too small, and it is also hastily bolted down to the surface plate ; the whole mass is then quickly immersed by a swing crane in a tank of water five feet deep, and hauled up and down till nearly cold ; the tires are not afterwards tempered. It is not indispensable that the whole tire should be of steel; but a dove-tail groove, turned out of the tire at the place where it bears most on the rail, and fitted with a band of steel, will suffice. This band may be put in in pieces, and the expedient appears to be the best way of repairing a worn tire; but particular care must be taken to attach these pieces very securely to the tire by rivets, else in the rapid revolution of the wheel the steel may be thrown out by the centrifugal force. In aid of such attachment the steel after being introduced is well hammered, which expands it sideways, until it fills the dove-tail groove. The tire is attached to the rim with rivets having counter-sunk heads, and the wheel is then fixed on its axle. The tire is turned somewhat conical, to facilitate the passage of the engine around curves,—the diameter of the outer wheel being virtually increased by the centrifugal force of the engine, and that of the inner wheel being correspondingly diminished, whereby the curve is passed without the resistance which would otherwise arise from the inequality of the spaces passed over by wheels of the same diameter fixed upon the same axle. The rails, moreover, are not set quite upright, but are slightly inclined inwards, in consequence of which the wheels must be either conical or slightly dished, to bear fairly upon the rails. One benefit of inclining the rails in this way, and coning the tires, is that the flange of the wheels is less liable to bear against the sides of the rail, and with the same view the flanges of all the wheels are made with large fillets in the corners. Wheels have been placed loose upon the axle, but they have less stability, and are not now much used. Much controversial ingenuity has been expended upon the question of the relative merits of the four and six wheeled engines; one party maintaining that four-wheeled engines are most unsafe, and the other that six-wheeled engines are unmechanical, and are more likely to occasion accidents. The four-wheeled engines, however, appear to have been charged with faults that do not really attach to them when properly constructed; for it by no means follows that if the axle of a four-wheeled engine breaks, or even altogether comes away, that the engine must fall down or run off the line; inasmuch as, if the engine be properly coupled with the tender, it has the tender to sustain it. It is obvious enough, that such a connection may be made between the tender and the engine, that either the fore or hind axle of the engine may be taken away, and yet the engine will not fall down, but will be kept up by the support which the tender affords; and the arguments hitherto paraded against the four-wheeled engines are, so far as regards the question of safety, nothing more than arguments against the existence of the suggested connection. It is no doubt the fact, that locomotive engines are now becoming too heavy to be capable of being borne on four wheels at high speeds without injury to the rails; but the objection of damage to the rails applies with at least equal force to most of the six-wheeled engines hitherto constructed, as in those engines the engineer has the power of putting nearly all the weight upon the driving wheels; and if the rail be wet or greasy, there is a great temptation to increase the bite of those wheels by screwing them down more firmly upon the rails. A greater strain is thus thrown upon the rail than can exist in the case of any equally heavy four-wheeled engine; and the engine is made very unsafe, as a pitching motion will inevitably be induced at high speeds, when an engine is thus poised upon the central driving wheels, and there will also be more of the rocking or sinuous motion. Stephenson makes his driving wheels without flanges, to facilitate the passage of the engine around curves; and if six-wheeled engines be made at all, it appears expedient to construct them in that manner; but instead of making enormously heavy six-wheeled engines, it appears the preferable alternative to use four-wheeled engines of a moderate weight, and to apply a sufficient number of them to a train, to enable it to reach the required velocity. To this arrangement there is no doubt the objection, that the expense of the propelling power is greater, as a small engine requires a driver and stoker for itself as well as a large engine; but by making the tender double, with one engine before it and another behind it, a single driver and a single stoker would suffice for the two engines. The starting handles of both engines might be brought to the middle of the tender, so that the engines might be started simultaneously, and be made to act in this respect like a single engine. This arrangement appears to me to be greatly preferable to that of making heavy six-wheeled engines, as the rail will be preserved from the injurious effects of excessive weight, and there will be less loss of power in contracting the blast pipe, when the fire and flue surface is increased by the addition of another engine. Tenders are now made larger than heretofore to obviate the necessity of so many coke and water stations; they should have glass windows all round them to shield the engine driver, and enable him during the worst winds and rains to keep a steady look out. Tenders may be put on any number of wheels, so that inconvenience is not likely to arise from their size and weight. The cranked axle of locomotives is always made of wrought-iron, with two cranks forged upon it towards the middle of its length, at a distance from each other answerable to the distance between the cylinders. Bosses are made on the axle for the wheels to be keyed upon, and bearings for the support of the framing. The axle is usually forged in two pieces, which are afterwards welded together. Sometimes the pieces for the cranks are put on separately, but the cranks so made are liable to give way. In engines with outside cylinders the axles are made straight—the crank pins being inserted in the naves of the wheels. The bearings to which the connecting rods are attached are made with very large fillets in the corners, so as to strengthen the axle in that part, and to obviate side play in the connecting rod. In engines which have been in use for some time, however, there is generally a good deal of end play in the bearings of the axles themselves, and this slackness contributes to make the oscillation of the engine more violent; but this evil may be remedied by making the bearings spheroidal, whereby end play becomes impossible. In every kind of locomotive it is very desirable that the length of the connecting rod should remain invariable, in spite of the wear of the brasses; for there is a danger of the piston striking against the cover of the cylinder if it be shortened, as the clearance is left as small as possible in order to economize steam. In some engines the strap encircling the crank pin is fixed immovably to the connecting rod by dove-tailed keys, and a bolt passes through the keys, rod, and strap, to prevent the dove-tail keys from working out. The brass is tightened by a gib and cutter, which is kept from working loose by three pinching screws and a cross pin or cutter through the point. The effect of this arrangement is to lengthen the rod, but at the cross-head end of the rod the elongation is neutralized by making the strap loose, so that, in tightening the brass, the rod is shortened by an amount equal to its elongation at the crank-pin end. The tightening here is also effected by a gib and cutter, which is kept from working loose by two pinching screws pressing on the side of the cutter. Both journals of the connecting rod are furnished with oil cups, having a small tube in the centre with siphon wicks. The connecting rod is a thick flat bar, with its edges rounded.

172. *Q.*—Will you explain in what manner the joints of an engine are made?

A.—Rust joints are not now much used in engines of any kind, yet it is necessary that the engineer should be acquainted with the manner of their formation. One ounce of sal-ammoniac in powder is mingled with 18 ounces, or a pound, of borings of cast-iron, and a sufficiency of water is added to wet the mixture thoroughly, which should be done some hours before it is wanted for use. Some persons add about half an onnce of flowers of brimstone to the above proportions, and a little sludge from the grindstone trough. This cement is caulked into the joints with a caulking iron, about three-quarters of an inch wide, and one quarter of an inch thick, and after the caulking is finished the bolts of the joints may be tried to see if they cannot be further tightened. The skin of the iron must, in all cases, be broken where the rust joint is to be made; and, if the place be greasy, the surface must be well rubbed over with nitric acid, and then washed with water, till no grease remains. The oil about engines has a tendency to damage rust joints by recovering the oxide. Coppersmiths staunch the edges of their plates and rivets by means of a cement, formed of pounded quicklime, with serum of blood, or white of egg; and in copper boilers such a substance may be useful in stopping the impalpable leaks which sometimes occur, though Roman cement appears to be nearly as effectual.

173. *Q.*—Will you explain the method of case-hardening the parts of engines?

A.—The most common plan for case-hardening consists in the insertion of the articles to be operated upon among horn or leather cuttings, bone dust, or animal charcoal, in an iron box provided with a tight lid, which is then put into a furnace for a period answerable to the depth of steel required. In some cases the plan pursued by the gunsmith may be employed with convenience. The article is inserted in a sheet-iron case amid bone dust, often not burned; the lid of the box is tied on with wire, and the joint luted with clay; the box is heated to redness as quickly as possible, and kept half an hour at a uniform heat: its contents are then suddenly immersed in cold water. The more unwieldy portions of an engine may be case-hardened by prussiate of potash—a salt made from animal substances, composed of two atoms of carbon and one of nitrogen, and which operates on the same principle as the charcoal. The iron is heated in the fire to a dull red heat, and the salt is either sprinkled upon it or rubbed on in a lump, or the iron is rubbed in the salt in powder. The iron is then returned to the fire for a few minutes, and finally immersed in water. By some persons the salt is supposed to act unequally, as if there were greasy spots upon the iron which the salt refused to touch, and the effect under any circumstances is exceedingly superficial; nevertheless, upon all parts not exposed to wear, a sufficient coating of steel may be obtained by this process. In the malleable iron work of engines scrap-iron has long been used, and considered preferable to other kinds; but if the parts are to be case-hardened, as is now the usual practice, the use of scrap-iron is to be reprehended, as it is almost sure to make the parts twist in the case-hardening process. In case-hardening, iron absorbs carbon, which causes it to swell; and as some kinds of iron have a greater capacity for carbon than other kinds, in case-hardening they will swell more, and any such unequal enlargement in the constituent portions of a piece of iron will cause it to change its figure. In some cases, case-hardening has caused such a twisting of the parts of an engine that they could not afterwards be fitted together; it is preferable, therefore, to make such parts as are to be case-hardened to any considerable depth of Lowmoor iron, which being homogeneous will absorb carbon equally, and will not twist.

174. *Q.*—What is the composition of the brass used in the construction of engines?

A.—The brass bearings of an engine are composed principally of copper and tin: the ordinary range of good yellow brass that files and turns well, is about 4½ to 9 oz. of zinc to the pound of copper. Brazing solders when stated in the order of their hardness are, —three parts copper and one part zinc (very hard), eight parts brass and one part zinc (hard), six parts brass, one part tin, and one part zinc (soft); a very common solder for iron, copper, and brass, consists of nearly equal parts of copper and zinc. Muntz's metal consists of forty parts zinc and sixty of copper; any proportions between the extremes of fifty parts of zinc and fifty parts copper, and thirty-seven zinc and sixty-three copper, will roll and work at a red heat, but forty zinc to sixty copper are the proportions preferred. Bell metal, such as is used for large bells, consists of 4½ oz. to 5 oz. of tin to the pound of copper; speculum metal consists of from 7½ oz. to 8½ oz. of tin to the pound of copper; tough brass for engine work, 1½ lb. tin, 1½ lb. zinc, and 10 lbs. copper; brass for heavy bearings, 2½ oz. tin, ½ oz. zinc, and 1 lb. copper. There is a great difference in the length of time brasses wear, as made by different manufacturers; but the difference arises as much from a different quantity of surface, as from a varying composition of the metal. Brasses should always be made strong and thick, as when thin they collapse upon the bearing, and increase the friction and the wear. Babbitt's patent lining metal for bushes has latterly been introduced in the bushes of locomotive axles and other machinery; it is composed of 1 lb. of copper, 1 lb. regulus of antimony, and 10 lbs. of tin, or other similar proportions, the presence of tin being the only material condition. The copper is first melted, then the antimony is added, with a small portion of tin; charcoal being strewed over the surface of the metal in the crucible to prevent oxidation. The bush or article to be lined having been cast with a recess for the soft metal, is to be fitted to an iron mould, formed of the shape and size of the bearing or journal, allowing a little in size for the shrinkage. Drill a hole for the reception of the soft metal, say $\frac{1}{2}$ to $\frac{3}{4}$ in. diameter, wash the parts not to be tinned with a clay wash to prevent the adhesion of the tin, wet the part to be tinned with alcohol, and sprinkle fine sal-ammoniac upon it; heat the article until fumes arise from the ammonia, and immerse it in a kettle of Banca tin, care being taken to prevent oxidation. When sufficiently tinned the bush should be soaked in water, to take off any particles of ammonia that may remain upon it, as the ammonia would cause the metal to blow. Wash with pipe clay and dry; then heat the bush to the melting point of tin, wipe it clean, and pour in the metal, giving it sufficient head as it cools; the bush should then be scoured with fine sand to take off any dirt that may remain upon it, and it is then fit for use. This metal wears for a longer time than ordinary gun metal, and its use is attended with very little friction. If the bearing heats, however, from the stopping of the oil hole or otherwise, the metal will be melted out. A metallic grease, containing particles of tin in the state of an impalpable powder, would probably be preferable to the lining of metal just described.

175. *Q.*—Have you any information to offer relative to the lubrication of engine bearings?

A.—A very useful species of oil cup is now employed in a number of steam vessels, and which, it is said, accomplishes a considerable saving of oil, at the same time that it more effectually lubricates the bearings. A ratchet wheel is fixed upon a little shaft which passes through the side of the oil cup, and is put into slow revolution by a pendulum attached to its outside, and in revolving it lifts up little buckets of oil, and empties them down a funnel upon the centre of the bearing. Instead of buckets a few short pieces of wire are sometimes hung on the internal revolving wheel, the drops of oil which adhere on rising from the liquid being deposited upon a high part set upon the funnel, and which, in their revolution, the hanging wires touch. By this plan, however, the oil is not well supplied at slow speeds, as the drops fall before the wires are in the proper position for feeding the journal. Another lubricator consists of a cock or plug inserted in the neck of the oil cup, and set in revolution by a pendulum and ratchet wheel, or any other means. There is a small cavity in one side of the plug which is filled with oil when that side is uppermost, and delivers the oil through the bottom pipe when it comes opposite to it. In some cases bearings heat from the existence of a cruciform groove on the top brass for the distribution of the oil, the effect of which is to leave the top of the bearings dry. In the case of revolving journals the plan of cutting a cruciform channel for the distribution of the oil does not do much damage; but in other cases, as in beam journals, for instance, it is most injurious, and the brasses cannot wear well wherever the plan is pursued. The right way is to make a horizontal groove along the brass where it meets the upper surface of the bearing, so that the oil may be well deposited on the highest point of the journal, leaving the force of gravity to send it downwards. This channel should, of course, stop short a small distance from each flange of the brass; otherwise the oil would run out at the ends.

176. *Q.*—Will you explain the operation of erecting engines in the workshop?

A.—In beginning the erection of side lever marine engines in the workshop, the first step is to level the bed plate lengthways and across, and strike a line up the centre, as near as possible in the middle, which indent with a chisel in various places, so that it may at any time be easily found again. Strike another line at right angles with this, either at the cylinder or crank centre, by raising a perpendicular in the usual manner. Lay the other sole plate alongside at the right distance, and strike a line at the cylinder or crank centre of it also, shifting either sole plate a little endways until these two transverse lines come into the same line, which may be ascertained by applying a straight edge across the two sole plates. Strike the rest of the centres across, and drive a pin into each corner of each sole plate, which file down level, so as to serve for points of reference at any future stage; next, try the cylinder, or plumb it on the inside roughly, and see how it is for height, in order to ascertain whether much will be required to be chipped off the bottom, or whether more requires to be chipped off the one side than the other. Chip the cylinder bottom fair; set it in its place, plumb the cylinder very carefully with a straight edge and silk thread, and scribe it so as to bring the cylinder mouth to the right height, then chip the sole plate to suit that height. The cylinder must then be tried on again, and the parts filed wherever they bear hard, until the whole surface is well fitted. Next chip the place for the framing; set up the framing, and scribe the horizontal part of the jaw with the scriber used for the bottom of the cylinder, the upright part being set to suit the shaft centres, and the angular flange of cylinder, where the stay is attached, having been previously chipped plumb and level. The stake wedges with which the framing is set up preparatorily to the operation of scribing, must be set so as to support equally the superincumbent weight, else the framing will spring from resting unequally, and it will be altogether impossible to fit it well. These directions obviously refer exclusively to the old description of side lever engine with cast-iron framing; but there is more art in erecting an engine of that kind with accuracy, than in erecting one of the direct action engines, where it is chiefly turned or bored surfaces that have to be dealt with. It will be proper, however, to describe the method pursued in erecting oscillating engines. The columns here are of wrought-iron, and in the case of small engines there is a template made of wood and sheet-iron, in which the holes are set in the proper positions, by which the upper and lower frames are adjusted; but in the case of large engines, the holes are set off by means of trammels. The holes for the reception of the columns are cast in the frames, and are recessed out internally: the bosses encircling the holes are made quite level across, and made very true with a face plate, and the pillars which have been turned to a gauge are then inserted. The top frame is next put on, and must bear upon the collars of the columns so evenly, that one of the columns will not be bound by it harder than another. If this point be not attained, the surfaces must be further scraped, until a perfect fit is established. The whole of the bearings in the best oscillating engines are fitted by means of scraping, and on no other mode of fitting can the same reliance be placed for exactitude. In fixing the positions of the centres in side lever engines, it appears to be the most convenient way to begin with the main centre. The height of the centre of the cross head at half stroke above the plane of the main centre, is fixed by the drawing of the engine, which gives the distance from the centre of cross head at half stroke to the flange of the cylinder; and from thence it is easy to find the perpendicular distance from the cylinder flange to the plane of the main centre, merely by putting a straight edge along level, from the position of the main centre to the cylinder, and measuring from the cylinder flange down to it, raising or lowering the straight edge until it rests at the proper measurement. The main centre is in that plane, and the fore and aft position is to be found by plumbing up from the centre line on the sole plate. To find the paddle shaft centre, plumb up from the centre line marked on the edge of the sole plate, and on this line lay off from the plane of the main centre the length of the connecting rod, if that length be already fixed, or otherwise the height fixed in the drawing of the paddle shaft above the main centre. To fix the centre for the parallel motion shaft when the parallel bars are connected with the cross head, lay off upon the plane of main centre the length of the parallel bar from the centre of the cylinder; deduct the length of the radius crank, and plumb up the central line of motion shaft; lay off on this line, measuring from the plane of main centre, the length of the side rod; this gives the centre of parallel motion shaft when the radius bars join the cross head, as is the preferable practice where parallel motions are used. The length of the connecting rod is the distance from the centre of the beam when level, or the plane of the main centre, to the centre of the paddle shaft. The length of the side rods is the distance from the centre line of the beam when level, to the centre of the cross head when the piston is at half stroke. The length of the radius rods of the parallel motion is the distance from the point of attachment on the cross head or side rod, when the piston is at half stroke, to the extremity of the radius crank, when the crank is horizontal; or in engines with the parallel motion attached to the cross head, it is the distance from the centre of the pin of the radius crank when horizontal to the centre of the cylinder. Having fixed the centre of the parallel motion shaft in the manner just described, it only remains to put the parts together when the motion is attached to the cross head; but when the motion is attached to the side rod, the end of the parallel bar must not move in a perpendicular line, but in an arc, the versed sine of which bears the same ratio to that of the side lever, that the distance from the top of the side rod to the point of attachment bears to the total length of the side rod. The parallel motion when put in its place should be tested by raising and lowering the piston by means of the crane: first, set the beams level, and shift in or out the motion shaft plummer blocks or bearings, until the piston rod is upright. Then move the piston to the two extremes of its motion; if at both ends the cross head is thrown too much out, the stud in the beam to which the motion side rod is attached is too far out, and must be shifted nearer to the main centre; if at the extremities the cross head is thrown too far in, the stud in the beam is not out far enough. If the cross head be thrown in at the one end, and out equally at the other, the fault is in the motion side rod, which must be lengthened or shortened to remedy the defect.

177. *Q.*—Will you explain how the slide valve of an engine is set?

A.—Place the crank in the position corresponding to the end of the stroke, which can easily be done in the shop with a level or plumb line; but in a steam vessel another method becomes necessary. Draw the transverse centre line, answering to the centre line of the crank shaft, on the sole plate of the engine, or on the cylinder mouth if the engine be direct action; describe a circle of the diameter of the crank pin upon the large eye of the crank, and mark off on either side of the transverse centre line a distance equal to the semidiameter of the crank pin. From the point thus found, stretch a line to the edge of the circle described on the large eye of the crank, and bring round the crank shaft till the crank pin touches the stretched line; the crank may thus be set at either end of its stroke. When the crank is thus placed at the end of the stroke, the valve must be adjusted so as to have the amount of lead, or opening on the steam side, which it is intended to give at the beginning of the stroke; the eccentric must then be turned round upon the shaft until the notch in the eccentric rod comes opposite the pin on the valve lever, and falls into gear; mark upon the shaft the situation of the eccentric, and put on the catches in the usual way. The same process must be repeated for going astern, shifting round the eccentric to the opposite side of the shaft, until the rod again falls into gear. In setting valves, regard must be had to the kind of engine, the arrangement of the levers, and the kind of valve employed. In setting the valves of locomotives a similar method is adopted: place the crank in the position answerable to the end of the stroke of the piston, and draw a straight line, representing the centre line of the cylinder, through the centres of the crank shaft and crank pin. From the centre of the shaft describe a circle with the diameter equal to the throw of the valve, another to represent the crank shaft, and a third circle to represent the path of the crank pin. From the centre of the crank shaft, draw a line perpendicular to the centre line of the cylinder and crank shaft, and draw another perpendicular at a distance from the first equal to the amount of the lap and the lead of the valve: the points in which this line intersects the circle of the eccentric are the points in which the centre of the eccentric should be placed for the forward and reverse motions. When the eccentric rod is attached directly to the valve, the radius of the eccentric, which precedes the crank in its revolution, forms with the crank an obtuse angle; but when by the intervention of levers the valve has a motion opposed to that of the eccentric rod, the angle contained by the crank and the radius of the eccentric must be acute, and the eccentric must follow the crank: in other words, with a direct attachment to the valve the eccentric is set *more* than one-fourth of a revolution in advance of the crank, and with an indirect attachment the eccentric is set *less* than one-fourth of a circle behind the crank. If the valve were without lead or lap the eccentric would be exactly one-fourth of a circle in advance of the crank or behind the crank, according to the nature of the valve connection; but as the valve would thus cover the port by the amount of the lap and lead, the eccentric must be set forward so as to open the port to the extent of the lead, and this is effected by the plan just described. In working locomotives the eccentrics sometimes shift upon the shaft, in which case they may be easily refixed by setting the valve open the amount of the lead, setting the crank at the end of the stroke, and bringing round the eccentric upon the shaft till the eccentric rod gears with the valve. It would often be troublesome in practice to get access to the valve for the purpose of setting it, and this may be dispensed with if the amount of lap on the valve and the length of the eccentric rod be known. To this end draw upon a board two straight lines at right angles to one another, and from their point of intersection as a centre describe two circles, one representing the circle of the eccentric, the other the crank shaft ; draw a straight line parallel to one of the diameters, and distant from it the amount of the lap and the lead ; the points in which this parallel intersects the circle of the eccentric are the positions of the forward and backward eccentrics. Through these points draw radial lines from the centre of the circle, and mark the intersections of these lines with the circle of the crank shaft; measure with a pair of compasses the chord of the arc intercepted between either of these intersections and the diameter which is at right angles with the crank; and the diameters being first marked on the shaft itself, it will follow that by transferring with the compasses the distance found in the diagram, and marking the point, the position of the eccentric will be fixed without difficulty.

178. *Q.*—Will you explain the method of putting engines into a steam vessel ?

A.—As an illustration of this operation it may be advisable to take the case of a side lever engine, and the method of proceeding is as follows :—First measure across from the inside of paddle bearers to the centre of the ship, to make sure that the central line, running in a fore and aft direction on the deck or beams, usually drawn by the carpenter, is really in the centre. Stretch a line across between the paddle bearers in the direction of the shaft: to this line in the centre of the ship where the fore and aft mark has been made, apply a square with arms six or eight feet long, and bring a line stretched perpendicularly from the deck to the keelson, accurately to the edge of the square; the lower point of the line where it touches the keelson will be immediately beneath the marks made upon the deck. If this point does not come in the centre of the keelson, it will be better to shift it a little, so as to bring it to the centre, altering the mark upon the deck correspondingly, provided either paddle shaft will admit of this being done—one of the paddle brackets being packed behind with wood, to give it an additional projection from the side of the paddle bearer. Continue the line fore and aft upon the keelson as nearly as can be judged in the centre of the ship; stretch another line fore and aft through the mark upon the deck, and look it out of winding with the line upon the keelson. Fix upon any two points equally distant from the centre, in the line stretched transversely in the direction of the shaft; and from those points as centres, and with any convenient radius, sweep across the fore and aft line to see that the two are at right angles; and, if not, shift the transverse line a little to make them so. From the transverse line next let fall a line upon each outside keelson, bringing the edge of the square to the line, the other edge resting on the keelson. A point will thus be got on each outside keelson perpendicularly beneath the transverse line running in the direction of the shaft, and a line drawn between those two points will be directly below the shaft. To this line the line of the shaft marked on the sole plate has to be brought, care being taken, at the same time, that the right distance is preserved between the fore and aft line upon the sole plate, and the fore and aft line upon the central keelson. Before any part of the machinery is put in, the keelsons should be dubbed fair and straight, and be looked out of winding by means of two straight edges. The art of placing engines in a ship is more a piece of plain common sense than any other feat in engineering, and every man of intelligence may easily settle a method of procedure for himself. Plumb lines and spirit levels, it is obvious, cannot be employed on board a vessel, and the problem consists in so placing the sole plates, without these aids, that the paddle shaft will not stand awry across the vessel, nor be carried forward beyond its place by the framing shouldering up more than was expected. As a plumb line cannot be used, recourse must be had to a square; and it will signify nothing at what angle with the deck the keelsons run, so long as the line of the shaft across the keelsons is squared down from the shaft centre. The sole plates being fixed, there is no difficulty in setting the other parts of the engine in their proper places upon them. The paddle wheels must be hung from the top of the paddle box to enable the shaft to be rove through them, and the cross stays between the engines should be fixed in when the vessel is afloat. To try whether the shafts are in a line, turn the paddle wheels, and try if the distance between the cranks is the same at the upper and under end, and the two horizontal centres; if not, move the end of the paddle shaft up or down, backwards or forwards, until the distance between the cranks at all the four centres is the same.

179. *Q.*—In what manner are the engines of a steam-vessel secured to the hull?

A.—The engines of a steamer are secured to the hull by means of bolts called holding-down bolts; and in most steam-vessels a good deal of trouble is caused by these bolts, which are generally made of iron. Sometimes they go through the bottom of the ship, and at other times they merely go through the keelson,—a recess being made in the floor or timbers to admit of the introduction of a nut. The iron, however, wears rapidly away in both cases, even though the bolts are tinned; and it has been found the preferable method to make such of the bolts as pass through the bottom, or enter the bilge, of Muntz's metal, or of copper. In a side lever engine, four Muntz's metal bolts may be put through the bottom at the crank end of the framing of each engine, four more at the main centre, and four more at the cylinder, making twelve through-bolts to each engine; and it is more convenient to make these bolts with a nut at each end, as in that case the bolts may be dropped down from the inside, and the necessity is obviated of putting the vessel on very high blocks in the dock, in order to give room to put the bolts up from the bottom. The remainder of the holding-down bolts may be of iron, and may, by means of a square neck, be screwed into the timber of the keelsons as wood screws—the upper part being furnished with a nut which may be screwed down upon the sole plate, so soon as the wood screw portion is in its place. If the cylinder be a fixed one it should be bolted down to the sole plate by as many bolts as are employed to attach the cylinder cover, and they should be of copper or brass, in any situation that is not easily accessible. In well formed bolts, the spiral groove penetrates about one-twelfth of the diameter of the cylinder round which it winds, so that the diameter of the solid cylinder which remains is five-sixths of the diameter over the thread. If the strain to which iron may be safely subjected in machinery is one-fifteenth of its utmost strength, or 4000 lbs. on the square inch, then 2180 lbs. may be sustained by a screw an inch in diameter, at the outside of the threads. The strength of the holding-down bolts may easily be computed, when the elevating force of the piston or main centre is known, but it is expedient very much to exceed this strength in practice, on account of the elasticity of the keelsons, the liability to corrosion, and other causes. It is difficult to fix engines effectually which have once begun to work in the ship, for in time the surface of the keelsons on which the engines bear becomes worn uneven, and the engines necessarily rock upon it. As a general rule, the bolts attaching the engines to the keelsons are too few and of too large a diameter: it would be preferable to have smaller bolts, and a greater number of them. In addition to the bolts going through the keelsons, or the vessel's bottom, there should be a large number of wood screws securing the sole plate to the keelson, and a large number of bolts securing the various parts of the engine to the sole plate. In iron vessels, holding-down bolts passing through the bottom are not expedient; and there the engine has merely to be secured to the iron plate of the keelsons, which are made hollow, to admit of a more effectual attachment.

180. *Q.*—What are the most important of the points which suggest themselves to you, in connection with the management of marine engines?

A.—The attendants upon engines should prepare themselves for any casualty that may arise, by considering possible cases of derangement, and deciding in what way they would act should certain accidents occur. The course to be pursued must have reference to particular engines, and no general rules can therefore be given, but every marine engineer should be prepared with the measures to be pursued in the emergencies in which he may be called upon to act, and where every thing may depend upon his energy and decision. In some cases of collision, the funnel is carried away and lost overboard, and such cases are among the most difficult for which a remedy can be sought. If flame come out of the chimney when the funnel is knocked away, so as to incur the risk of setting the ship on fire, the uptake of the boiler must be covered over with an iron plate, or be sufficiently covered to prevent such injury. A temporary chimney must then be made of such materials as are on board the ship. If there are bricks and clay or lime on board, a square chimney may be built with them; or if there be sheet-iron plates on board, a square chimney may be constructed of them. In the absence of such materials, the awning stanchions may be set up round the chimney, and chain rove in through among them in the manner of wicker work, so as to make an iron wicker chimney, which may then be plastered outside with wet ashes mixed with clay, flour, or any other material that will give the ashes cohesion. War steamers should carry short spare funnels, which may easily be set up should the original funnel be shot away; and if a jet of steam be let into the chimney, a very short and small funnel will suffice for the purpose of draught. If the crank pin breaks, the other engine must be worked with the one wheel. It will sometimes happen, when there is much lead upon the slide valve, that the single engine, on being started, cannot be got to turn the centre, if there be a strong opposing wind and sea; the piston going up to near the end of the stroke, and then coming down again without the crank being able to turn the centre. In such cases, it will be necessary to turn the vessel's head sufficiently from the wind to enable some sail to be set; and if once there is weigh got upon the vessel the engine will begin to work properly, and will continue to do so though the vessel be put head to wind as before. If the eccentric catches, or hoops break or come off, and the damage cannot readily be repaired, the valve may be worked by attaching the end of the starting handle to any convenient part of the other engine, or to some part in connection with the connecting rod of the same engine. In side lever engines, with the starting bar hanging from the top of the diagonal stay, as is a very common arrangement, the valve might be wrought by leading a rope from the side lever of the other engine through blocks, so as to give a horizontal pull to the hanging starting bar, and the bar could be brought back by a weight. Another plan would be, to lash a piece of wood to the cross-tail butt of the damaged engine, so as to obtain a sufficient throw for working the valve, and then to lead a piece of wood or iron, from a suitable point in the piece of wood attached to the cross-tail, to the starting handle, whereby the valve would receive its proper motion. If the shafts or cranks break, the engine may nevertheless be worked with moderate pressure to bring the vessel into port; but if the crank be very bad, it will be expedient to fit strong blocks of wood under the ends of the side levers, or other suitable part, to prevent the cylinder bottom or cover from being knocked out, should the damaged part give way. The same remark is applicable when flaws are discovered in any of the main parts of the engine, whether they be malleable or cast-iron; but they must be carefully watched, so that the engines may be stopped if the crack is extending further. Should fracture occur, the first thing obviously to be done is to throw the engines out of gear, and should there be much weigh on the vessel, the steam should at once be thrown on the reverse side of the piston, so as to counteract the pressure of the paddle wheel.

181. *Q.*—What are the chief duties of the engine driver of a locomotive?

A.—The engineer of a locomotive should constantly be upon the foot-board of the engine, so that the regulator, the whistle, or the reversing handle may be used instantly, if necessary; he must see that the level of the water in the boiler is duly maintained, and that the steam is kept at a uniform pressure. In feeding the boilers with water, and the furnaces with fuel, a good deal of care and some tact is necessary, as irregularity in the production of steam will often occasion priming, even though the water be maintained at a uniform level; and an excess of water will of itself occasion priming, while a deficiency is a source of obvious danger. The engine is generally furnished with three gauge cocks, and water should always come out of the second gauge cock and steam out of the top one when the engine is running; but when the engine is at rest, the water in the boiler is rather lower than when in motion, so that when the engine is at rest, the water will be high enough if it just reaches to the middle gauge cock. The boiler should be well filled with water on approaching a station, as there is then steam to spare, and additional water cannot be conveniently supplied when the engine is stationary. The furnace should be fed with small quantities of fuel at a time, and the feed should be turned off just before a fresh supply of fuel is introduced. The regulator may, at the same time, be partially closed; and, if the blast pipe be a variable one, it will be expedient to open it widely while the fuel is being introduced, to check the rush of air in through the furnace door, and then to contract it very much so soon as the furnace door is closed, in order to recover the fire quickly. The proper thickness of coke upon the grate depends upon the intensity of the draught: but in heavily loaded engines it is usually kept up to the bottom of the fire door. Care, however, must be taken that the coke does not reach up to the bottom row of tubes so as to choke them up. The fuel is usually disposed on the grate like a vault; and if the fire box be a square one, it is heaped high in the corners, the better to maintain the combustion. In starting from a station, and also in as ending inclined planes, the feed water is generally shut off; and therefore, before stopping or ascending inclined planes, the boiler should be well filled up with water. In descending inclined planes an extra supply of water may be introduced into the boiler, and the fire may be fed, as there is at such times a superfluity of steam. In descending inclined planes the regulator must be partially closed, and it should be entirely closed if the plane be very steep. The same precaution should be observed in the case of curves, or rough places on the line, and in passing over points or crossings. To ascertain whether the pumps are acting well, the pet cock, which is a small cock opening into the pump, must be turned, and if any of the valves stick they will sometimes be induced to act again by working with the pet cock open, or alternately open and shut. Should the defect arise from a leakage of steam into the pump, which prevents the pump from drawing, the pet cock remedies the evil by permitting the steam to escape. Should priming occur from the water in the boiler being dirty, a portion of it may be blown out; and should there be much boiling down through the glass-gauge tube, the stop cock may be partially closed. The water should be wholly blown out of locomotive boilers three times a week, and at those times two mud hole doors at opposite corners of the boiler should be opened, and the boiler be washed internally by means of a hose. On approaching a station the regulator should be gradually closed, and it should be completely shut about half a mile from the station—if the train be a very heavy one; the train may then be brought to rest by means of the breaks. Too much reliance, however, must not be put upon the breaks, as they sometimes give way, and in frosty weather are nearly inoperative. In cases of urgency the steam may be thrown upon the reverse side of the piston, but it is desirable to obviate this necessity as far as possible. At terminal stations the steam should be shut off earlier than at roadside stations, as a collision will take place at terminal stations if the train overshoots the place where it ought to stop. There should always be a good supply of water when the engine stops; but the fire may be suffered gradually to burn low towards the conclusion of the journey. So soon as the engine stops it should be wiped down, and be then carefully examined: the brasses should be tried, to see whether they are slack or have been heating; and, by the application of a gauge, it should be ascertained occasionally whether the wheels are square on their axles, and whether the axles have end play, which should be prevented. The stuffing boxes must be tightened, and the valve gear examined, and the eccentrics be occasionally looked at to see that they have not shifted on their axles, though this defect will be generally intimated by the irregular beating of the engines. The tubes should also be examined and cleaned out, and the ashes emptied out of the smoke box through the small ash door at the end. If the engine be a six-wheeled one, it will be liable to pitch and oscillate if too much weight be thrown upon the driving wheels; and where such faults are found to exist, the weight upon the driving wheels should be diminished. The practice of blowing off the boiler by the steam, as is always done in marine boilers, should not be permitted as a general rule in locomotive boilers, when the tubes are of brass and the fire box of copper; but when the tubes and fire boxes are of iron, there will not be an equal risk of injury. Before starting on a journey, the engine man should take a summary glance beneath the engine—but before doing so he ought to assure himself that no other engine is coming up at the time. The regulator, when the engine is standing, should be closed and locked, and the eccentric rod be fixed out of gear, and the tender break screwed down; the cocks of the oil vessels should at the same time be shut, but should all be opened a short time before the train starts. When a tube bursts, a wooden or iron plug must be driven into each end of it, and if the water or steam be rushing out so fiercely that the exact position of the imperfection cannot be discovered, it will be advisable to diminish the pressure by increasing the supply of feed water. Should the leak be so great that the level ot the water in the boiler cannot be maintained, it will be expedient to drop the bars and quench the fire, so as to preserve the tubes and fire box from injury. Should the wooden casing of the boiler catch fire, it may be extinguished by throwing a few buckets of water upon it, or, if the engine is at a station, it may be brought under the water crane. Should the piston rod or connecting rod break, or the cutters fall out or be clipped off—as sometimes happens to the piston cutter when the engine is suddenly reversed upon a heavy train—the parts should be disconnected, if the connection cannot be restored, so as to enable one engine to work; and of course the valve of the faulty engine must be kept closed. If one engine has not power enough to enable the train to proceed with the blast pipe full open, the engine may perhaps be able to take on a part of the carriages, or it may run on by itself to fetch assistance. The same course must be pursued if any of the valve gearing becomes deranged, and the defects cannot be rectified upon the spot. To economize fuel, the variable expansion gear, if the engine has one, should be adjusted to the load, and the blast pipe should be worked with the least possible contraction; and at stations the damper should be closed to prevent the dissipation of heat.

182. *Q.*—Having now given a general outline of the principles and details of construction involved in the production of steam engines, according to the existing system, can you give any suggestion of material improvement?

A.—For all land purposes, and for some marine purposes, engines on the high speed principle (described in pages 68 and 69) should be adopted; and in locomotive engines much change is necessary, as the existing locomotive engine intended for high speeds upon the narrow gauge is a very defective machine. It is a monstrous thing that, in the present age of mechanical proficiency, half the engine power should be dissipated upon the blast pipe, as in high speeds appears to be the case; and the first indication of improvement is to obviate this loss by enlarging the grate surface and the area of the tubes. This may be most conveniently accomplished by the introduction of vertical tubes, inserted in short barrels, each furnished with a chimney, and set upon a square fire-box extending the whole length of the engine ; the axle of the driving wheel may be introduced above the fire-box, and between two of the barrels, whereby the engine will be kept so near the ground as to obviate the liability of being top-heavy. If the two cylinders, which may be oscillating, be set upon the top of the fire-box, at the angle of 45° with each other, but in the same vertical plane, and if both piston rods be coupled to the same crank, the shaft of which extends to each side of the carriage, where it is provided with other cranks operating by means of horizontal rods upon pins in the driving wheel, the highest velocities will be attainable without any sinuous or rocking motion ; and the whole of the machinery being above the boiler, will be accessible while the train is running, and will not be exposed so much to ashes and dust. With such a construction, indeed, it would be expedient to enclose the whole of the machinery in a carriage or house with glass windows round it. Every locomotive should be furnished with efficient expansion gear of some kind or other ; and it is very desirable that the fire should be fed by some self-acting arrangement, which will make it unnecessary to open the furnace-door frequently. The use of sediment collectors in locomotive boilers is also expedient, as, if judiciously applied, they will effectually prevent the formation of scale upon the tubes, and will also operate as an antidote to priming, in many cases. The form of collector best adapted to a locomotive boiler will depend, in a great measure, upon the peculiar structure of the boiler; but generally any form will answer which communicates with the water level, and contains water within it in a tranquil state.

TABLE 1.—*Page* 277.

Table of Nominal Horse Power of Low Pressure Engines.

THE table on the opposite page contains the dimensions of cylinder corresponding with any given power in a low pressure engine. The table is constructed according to the rule given in page 50.

To find the nominal power of a low pressure engine, of 20 inches diameter of cylinder, and 2½ feet stroke:—Find the diameter in the left-hand column, and in the same horizontal line with it, and in the vertical column headed 2½, will be found 11·55, the number of horses power of the engine.

The dimensions of cylinder corresponding to a given power may be found by a converse process.

Table of Nominal Horse Power of Low Pressure Engines.

Diam. of Cylinder in inches.	Length of Stroke in Feet.											
	1	1½	2	2½	3	3½	4	4½	5	5½	6	7
4	·34	·39	·43	·46	·49	·52	·54	·56	·58	·60	·62	·65
5	·53	·61	·67	·72	·76	·81	·84	·88	·91	·94	96	1·02
6	·76	·87	·96	1·04	1·10	1·16	1·22	1·26	1.31	1·35	1·39	1·47
7	1·04	1·19	1·31	1·41	1·50	1·58	1·65	1·72	1·78	1·84	1·89	1·99
8	1·36	1·56	1·72	1·85	1·96	2·07	2·16	2·25	2·33	2·40	2·47	2·60
9	1·72	1·97	2·17	2·34	2·49	2·62	2·74	2·84	2·95	3·04	3·13	3·30
10	2·13	2·44	2·68	2.89	3·07	3·23	3·38	3·51	3·64	3·76	3·87	4·07
11	2·57	2·95	3·24	3·49	3·77	3·91	4·15	4·25	4·40	4·54	4·68	4·92
12	3·06	3·51	3·86	4·16	4·42	4·65	4·86	5·06	5·24	5·41	5·57	5·86
13	3·60	4·12	4·53	4·88	5·19	5·46	5·64	5·94	6·15	6·35	6·53	6·88
14	4·17	4·77	5·25	5·66	6·01	6·33	6·62	6·88	7·13	7·36	7·58	7·98
15	4·77	5·48	6·03	6·50	6·90	7·27	7·60	7·90	8·19	8·45	8·70	9·16
16	5·45	6·23	6·86	7·39	7·86	8·27	8·65	8·99	9·31	9·61	9·90	10·42
17	6·15	7·04	7·75	8·35	8·86	9·34	9·76	10·15	10·52	10·85	11·17	11·76
18	6·89	7·89	8·68	9·36	9·94	10·47	10·94	11·38	11·79	12·17	12·53	13·19
19	7·68	8·79	9·68	10·42	11·17	11·66	12·19	12·68	13·13	13·56	13·96	14·69
20	8·51	9·74	10·72	11·55	12·27	12·92	13·51	14·05	14·55	15·02	15·46	16·28
22	10·30	11·79	12·97	13·98	14·85	15·63	16·62	17·30	17·65	18·18	18·71	19·70
24	12·26	14·03	15·44	16·63	17·67	18·61	19·45	20·23	20·95	21·63	22·27	23·44
26	14·39	16·46	18·12	19·52	20·75	21·84	22·56	23·75	24·6	25·39	26·14	27·51
28	16·68	19·09	21·02	22·64	24·06	25·33	26·48	27·54	28·52	29·44	30·31	31·90
30	19·15	21·92	24·13	25·99	27·62	29·07	30·40	31·61	32·74	33·80	34·80	36·63
32	21·79	24·96	27·51	29·57	31·42	33·08	34·59	35·97	37·26	38·46	39·59	41·68
34	24·60	28·16	30·99	33·39	35·44	37·34	39·04	40·60	42·06	43.41	44·69	47·05
36	27·57	31·56	34·74	37·42	39·77	41·87	43·77	45·52	47·15	48.67	50·11	52·75
38	30·72	35·17	38·71	41·69	44·66	46·64	48·77	50·72	52·54	54.23	55·83	58·78
40	34·04	38 97	42·89	46·20	49·10	51·69	54·04	56·20	58·21	60.09	61·86	65·12
42	37·53	42·96	47·29	50·94	54·13	56·98	59·58	61·96	64·18	66.25	68·21	71·78
44	41·19	47·15	51·90	55·91	59·38	62·54	66·46	68·00	70·44	72.71	74·85	78·79
46	45·02	51·54	56·72	61·10	64·88	68·19	71·43	74·33	76·69	79.47	81·81	86·12
48	49·02	56·11	61·76	66·53	70·70	74·42	77·82	80·94	83·83	86.53	89·08	93·78
50	53·19	60·89	67·02	72·19	76·71	80·76	84·44	87·82	90·96	93 89	96·65	101·7
52	57·55	65·86	72·48	78·08	83·00	87·35	90·25	94·98	98·40	101.55	104·5	110.0
54	62·04	71·02	78·17	84·20	89·48	94·20	98·49	102·4	106·1	109.5	112·7	118·7
56	66·72	76·38	84·07	90·55	96·23	101·30	105·9	110·1	114·1	117·8	121·2	127·6
58	71·58	81·93	90·18	97·14	103·2	108·6	113·6	118·2	122·4	126.3	129.9	136·7
60	76·60	87·68	96·50	103·9	110·4	116·3	121·6	126·4	131·0	135·2	139·2	146·5
62	81·79	93·62	103·04	111·0	117·96	124·18	129·81	135·03	139·86	144·37	148·6	156·7
64	87·15	99·84	110·0	118·3	125·7	132·3	138·3	143·9	149·0	153·82	158·4	166·7
66	92·68	106·1	116·8	125·8	133·6	140·7	147·3	153·0	158·5	163·6	168·4	177·3
68	98·40	112·6	123·9	133·6	141·8	149·4	156·2	162·4	168·2	173·6	178·8	188·2
70	104·26	119·3	131·3	141·5	150·4	158·3	165·5	172·1	178·2	184·0	189·4	199·4
72	110·30	126·2	139·0	149·7	159·1	167 4	175·1	182·1	188·6	194·7	200·4	211·0
74	116·5	133·4	146·8	158·1	167·9	176·7	185·4	192·4	199·2	205·7	211·6	223·4
76	122·9	140·7	154·8	166·8	178·6	186·6	195·0	202·9	210·1	216·9	223·3	235·1
78	129·4	148·2	163·1	175·6	186 7	196·5	205·4	212·1	221·4	228·5	235·2	247·6
80	136·2	155·8	171·6	184·8	196·4	206·7	216·1	224·8	232·8	240·4	247·4	260·5
82	143·0	163·8	180·2	194·2	206·2	217·3	226·9	237·8	244·6	252·5	260·0	273·8
84	150·1	171·8	189·1	203·8	216·5	227·9	238·3	247·8	256.7	265·0	272·8	287·1
86	157·4	180·1	198·2	213·6	227·0	237·8	247·4	258·2	269·1	277·8	286·0	301·0
88	164·8	188·6	207·6	223·6	237·5	250·2	261·6	272·0	281·7	290·8	299·4	315·2
90	172·3	197·3	217·1	233·9	248·6	261·7	273·6	284·5	294·7	304·2	313·2	329·7

Table 2.—*Page* 279.

Table of Nominal Horse Power of High Pressure Engines.

The table on the opposite page contains the dimensions of cylinder corresponding with any given power in a high pressure engine. The table is constructed according to the rule given in page 67, which is virtually the same rule as that for low pressure engines, with the exception of the element of pressure being taken at three times the value.

To find the nominal power of a high pressure engine of 20 inches diameter of cylinder, and 2½ feet stroke :— Find the diameter in the left-hand column, and in the same horizontal line with it, and in the vertical column headed 2½, will be found 34·65, the number of horses power of the engine.

The dimensions of cylinder corresponding to a given power may be found by the converse process from the table.

Table of Nominal Horse Power of High Pressure Engines.

Diam. of Cylinder in inches.	Length of Stroke in Feet.											
	1	1½	2	2½	3	3½	4	4½	5	5½	6	7
2	·25	·29	·32	·35	·37	·38	·40	·42	·44	·45	·46	·49
2½	·39	·45	·50	·54	·57	·60	·63	·66	·68	·70	·72	·76
3	·57	·65	·72	·78	·83	·87	·91	·95	·98	1·01	1·04	1·10
3½	·78	·89	·98	1·06	1·13	1·19	1·24	1·29	1·34	1·38	1·42	1·49
4	1·02	1·17	1·29	1·38	1 47	1·56	1·62	1·68	1·74	1·80	1·86	1·95
4½	1·29	1·48	1·63	1·75	1·86	1·96	2·05	2·13	2·21	2·28	2·35	2·47
5	1·59	1·83	2·01	2·16	2·28	2 43	2·52	2·64	2·73	2 82	2·88	3·06
5½	1·93	2·21	2·43	2·62	2·78	2·93	3·12	3·18	3·30	3·42	3·51	3·69
6	2·28	2·61	2·88	3·12	3·30	3·48	3·66	3·78	3·93	4·05	4·17	4·41
6½	2 69	3 09	3·39	3·66	3·90	4·08	4·23	4·44	4 62	4 77	4·89	5·16
7	3·12	3·57	3 93	4·23	4·50	4·74	4·95	5·16	5·34	5·52	5·67	5·97
7½	3·60	4·11	4·53	4·86	5·19	5·46	5·70	5·94	6·15	6·33	6·51	6·87
8	4·08	4·68	5·16	5·55	5·88	6 21	6·48	6·75	6·99	7·20	7·41	7·80
8½	4·62	5·28	5·82	6·27	6·63	6·99	7 32	7·62	7·89	8·13	8·37	8 82
9	5·16	5·91	6·51	7·02	7·47	7·86	8·22	8·52	8·85	9·12	9·39	9·90
9½	5·76	6·60	7·26	7·80	8·37	8·76	9·15	9·51	9·84	10·17	10·47	11·01
10	6·39	7·32	8·04	8·67	7·21	9·69	10·14	10·53	10·92	11·28	11·61	12·21
10½	7·05	8·04	8·88	9·54	10·14	10·68	11·16	11·61	12·03	12·42	12·78	13·47
11	7·71	8·85	9·72	10·47	11·31	11·73	12·45	12·75	13·20	13·62	14·04	14·76
11½	8·43	9·06	10·62	11·46	12·15	12·78	13·80	13·92	14·61	14·91	15·33	16·14
12	9·18	10·53	11·58	12·48	13·26	13·95	14·58	15·18	15·72	16·23	16·71	17 58
12½	9·96	11·40	12·57	13·53	14·37	15·15	15·84	16·47	17·04	17·58	18 12	19·08
13	10·80	12·36	13·59	14·64	15·57	16·35	16·92	17·82	18·45	19·05	19·59	20·64
13½	11·64	13·32	14·64	15·78	16·77	17·67	18·48	19 20	19·89	20·52	21·15	22·26
14	12·51	14·31	15·75	16 98	18·03	18·99	19·86	20·64	21·59	22·08	22·74	23·94
14½	13·41	15·56	16·92	18·21	19·35	20·37	21·30	22·14	22·95	23·70	24·39	25·62
15	14·31	16·44	18·09	19·50	20·70	21·81	22·80	23·70	24·57	25·35	26·10	27·48
16	16·35	18·69	20·58	22·17	23·58	24·81	25·95	26·97	27 93	28 83	29·70	31·26
17	18·45	21·12	23·25	25·05	26·58	28·02	29·28	30 45	31·56	3[illegible]·55	33 57	35 28
18	20·67	23·67	26·04	28·08	29 82	31·41	32·82	34·14	35·37	36·51	37·59	39 57
19	23·04	26·37	29·04	31·26	35·51	34·98	36·57	38·04	39·39	40·68	41·88	44·07
20	25·53	29 22	32·16	34·65	36·81	38·76	40·53	42·15	43·65	45·06	46·38	48·84
22	30·90	35·37	38·91	41·94	44·55	46·89	49·86	51·90	52·95	54·54	56·13	59·10
24	36·78	42·09	46·32	49·[illegible]9	53·01	55·83	58·35	60·69	62·85	64·89	66·81	70·32
26	43·17	49·38	54 36	58·56	62·25	65·52	67 68	7[illegible]·25	7[illegible]·8	76·17	78·42	82·53
28	50·04	57 27	63·06	67·92	72·18	75·99	79·44	82·62	85·56	8[illegible]·32	90·93	95·70
30	57·45	65·76	72·39	77·97	82·86	87·21	91·20	94·83	98·22	101·40	104·4	109 9
32	65·37	74 88	82·53	88·71	94 26	99·24	103·7	107·9	111·8	115·4	118·7	125·0
34	73·80	84 48	92·97	100·22	106·3	112·0	117·1	121·8	126·2	130·2	134·0	141·1
36	82·71	94·68	104·2	112·2	119·3	125·6	131·3	136·5	14[illegible]·4	146·0	150·3	158·2
38	92·16	105·5	116·1	125·0	134·0	136·9	146·3	152·1	157·6	162·7	167·5	176·3
40	1[illegible]2·1	116·9	12[illegible]·6	128·6	147·3	155·1	162·1	168·6	174·6	180·2	185·6	195·3
42	112·6	128·9	141·8	152·8	162·4	170·9	178·7	185·9	192·5	198·7	204·6	215·3
44	123·5	141·4	155·7	167·7	178·1	187·6	199·4	204·0	211·3	218·1	224·5	236·3
46	135·0	154·6	170·1	183·3	194·6	204·6	214·3	223·0	230·0	238·4	245·4	258·3
48	147·0	168·3	185·3	199·6	212·1	223·2	233·4	242·8	251·5	259·6	267·2	281·3
50	159·6	182·6	201·0	216·5	230·1	242·3	253·3	263·4	272·9	281·6	289·9	305·1
52	172·6	197·6	217·4	234 2	249·0	262·0	270·7	284·9	295·2	304·6	313·5	330·0
54	186·1	213·0	234·5	252·6	268·4	282·6	295·4	307·2	318·3	328·5	338·1	356·1
56	200·1	229·1	252·2	271·6	288·7	303·9	317·7	330·3	342·3	353·4	363·6	382·8
58	214·7	245·8	270·5	291·4	309·6	325 8	340·8	354·6	367·2	378·9	389·7	410·1
60	229·8	263·0	289·5	311·7	331·2	348·9	364·8	379·2	393·0	405·6	417·6	439·5

TABLE 3.—*Page* 281.

Expanded Steam.—Mean Pressure at different Densities and Rates of Expansion.

THE tables on the opposite page are intended to facilitate the computation of the power of an engine working expansively. The first column in each table contains the initial pressure of the steam in pounds; and the remaining columns contain the mean pressure of steam throughout the stroke, with the different degrees of expansion indicated at the top of the columns, and which express the portion of the stroke during which the steam acts expansively. Thus, for example, if steam be admitted to the cylinder at a pressure of 3 pounds per square inch, and be cut off within $\frac{1}{8}$th of the end of the stroke, the mean pressure during the whole stroke will be 2·96 pounds per square inch. In like manner, if steam, at the pressure of 3 pounds per square inch, were cut off after the piston had gone through $\frac{1}{8}$th of the stroke, leaving the steam to expand through the remaining $\frac{7}{8}$ths, the mean pressure during the whole stroke would be 1·154 pounds per square incn. Steam, however, of as little as 3 lbs. per square inch, is never used in working an engine, for this would be steam of 12 lbs. below the atmospheric pressure; and if steam of 3 lbs. *above* the atmospheric pressure be taken, this will be steam of 18 lbs. pressure upon the square inch; the mean pressure of which throughout the stroke, if cut off at the time the piston has gone through $\frac{1}{8}$th of the stroke, or has $\frac{7}{8}$ths of the stroke still to perform, is 6·927 lbs. per square inch, being nearly 8 lbs. below the atmospheric pressure.

EXPANDED STEAM.

Mean Pressure at Different Densities and Rates of Expansion.

The Columns headed 0 *contain the Initial Pressures in lbs., and the remaining columns contain the Mean Pressure in lbs., with different Grades of Expansion.*

Expansion by Eighths.

0	1/8	2/8	3/8	4/8	5/8	6/8	7/8
3	2·96	2·89	2·75	2·53	2·22	1·789	1·154
4	3·95	3·85	3·67	3·38	2·96	2·386	1·539
5	4·948	4·818	4·593	4·232	3·708	2·982	1·924
6	5·937	5·782	5·512	5·079	4 450	3·579	2·300
7	6·927	6·746	6·431	5·925	5·241	4·175	2·694
8	7·917	7·710	7·350	6·772	5·934	4·772	3 079
9	8·906	8·673	8·268	7·618	6·675	5·368	3·463
10	9·896	9·637	9·187	8·465	7·417	5·965	3·848
11	10·885	10·601	10·106	9·311	8·159	6·561	4·233
12	11·875	11·565	10·925	10·158	8·901	7·158	4·618
13	12 865	12·528	11·943	11·004	9·642	7·754	5·003
14	13·854	13·492	12·862	11 851	10·384	8·531	5·388
15	14·844	14·456	13·781	12·697	11·126	8·947	5·773
16	15·834	15·420	14·700	13·544	11·868	9·544	6·158
17	16·823	16·383	15·618	14·390	12·609	10·140	6·542
18	17·813	17·347	16·537	15·237	13·351	10·737	6·927
19	18·702	18·311	17·448	16·803	14·093	11·333	7·312
20	19·792	19·275	18·375	16·930	14·835	11·930	7·697
25	24·740	24·093	22·968	21·162	18·543	14·912	9·621
30	29·688	28·912	27·562	25·395	22·252	17·895	11·546
35	34·636	33·731	33·156	29·627	25·961	20·877	13·470
40	39·585	38·550	36·750	33·860	29·670	23·860	15·395
45	44·533	43·368	41·343	38 092	33·378	26·842	17·319
50	49·481	48·187	45·937	42·325	37·067	29·825	19·243

Expansion by Tenths.

0	1/10	2/10	3/10	4/10	5/10	6/10	7/10	8/10	9/10
3	2·980	2·930	2·830	2·710	2·539	2·299	1·981	1·568	0·990
4	3·974	3 913	3 780	3·614	3·386	3·065	2·642	2·087	1·320
5	4·968	4·892	4·725	4·518	4·232	3·832	3·303	2·609	1·651
6	5·961	5·870	5·670	5·421	5·079	4 598	3·963	3·130	1·981
7	6·955	6·848	6·615	6·325	5·925	5·364	4·624	3·652	2·311
8	7·948	7 827	7·560	7·228	6·772	6·131	5·284	4·174	6·641
9	8·942	8·805	8·505	8·132	7·618	6·897	5·945	4·696	2·971
10	9·936	9·784	9·450	9·036	8·465	7·664	6·606	5·218	3·302
11	10·929	10·762	10·395	9·939	9·311	8·430	7·266	5·739	3·632
12	11 923	11·740	11·340	10·843	10·158	9·196	7·927	6·261	3·962
13	12·856	12·719	12·285	11·746	10·994	9·963	8·587	6·783	4·292
14	13·910	13·967	13·230	12·650	11·851	10·729	9·248	7·305	4·622
15	14·904	14·676	14·175	13·554	12·697	11·496	9·909	7·827	4·953
16	15·897	15·654	15·120	14·457	13·544	12·262	10·569	8·348	5·283
17	16·891	16·632	16·065	15·361	14·051	13·028	11·230	8·870	5·613
18	17·884	17·611	17·010	16·264	15·237	13·795	11·890	9·392	5·944
19	18·878	18·589	17·955	17·168	16·083	14·561	12·551	9·914	6·273
20	19·872	19·568	18·900	18·072	16·930	15·328	13·212	10·436	6·600
25	24·840	24 460	23·625	22·590	21·162	19·160	16·515	13·040	8·255
30	29·808	29·352	28·350	27·108	25·395	22·992	19·818	15·654	9·960
35	34.776	34·244	33·075	31·626	29.627	26·824	23·121	18·263	11·557
40	39·744	39·136	37·800	36·144	33·860	30·656	26·224	20·872	13·208
45	44·912	44·028	42·525	40·662	38·092	34·888	29·727	23·481	14·859
50	49·680	48·920	47·250	45·180	42·325	38·320	33·030	26·090	16·510

TABLE 4.—*Page* 283.

TABLE 4 contains four small tables. The first gives the temperature and elastic force of steam, as experimentally determined by Dulong and Arago; the second gives the expansion of air by heat; the remaining tables have reference to the use of steam expansively, by means of *lap* on the valve.

The larger of these tables is intended to show the amount of cover required on the steam side of the valve to cut off the steam at different parts of the stroke. The first column contains different lengths of the stroke of the valve in inches, and the remaining columns are headed by fractions, indicating the degree of expansion. To cut off the steam at $\frac{1}{6}$th from the end of the stroke, the stroke of the valve being 17 inches:—In the column headed $\frac{1}{6}$, and in a line with 17, will be found 3·47, which is the required cover in inches, the valve being without lead. To give it $\frac{1}{4}$ inch of lead, subtract one-half the lead from the cover, and the remainder (3·345 inches) is the cover for the required expansion and lead. Subtract the cover from one-half the length of the valve stroke, and the remainder (5·155 inches) is the greatest breadth of port with that stroke.

The remaining table is to show at what parts of the stroke, under any given arrangement of slide valve, the exhausting ports close and open. The columns headed A contain the amount of cover on the exhausting side of the valve in parts of its stroke. In the column B, the figures represent the distance of the piston, in terms of the length of its stroke, from the end of its stroke, when the exhausting port before it is shut; and C shows similarly at what part of the stroke the exhausting port behind the piston is opened—the steam, in both cases, being cut off at $\frac{1}{3}$d from the end of the stroke. In like manner, in D E the steam is cut off at $\frac{7}{24}$ths from the end of the stroke; in F G at $\frac{1}{4}$th; in H K at $\frac{5}{24}$ths; in L M at $\frac{1}{6}$th; in N O at $\frac{1}{8}$th; in P Q at $\frac{1}{12}$th; in R S at $\frac{1}{24}$th. In each case, the left hand single column of each double column represents the distance of the piston from the end of its stroke, when the exhausting port before it is shut; and the right hand single column when the exhausting port behind it is opened.

Let the stroke of the piston be 6 feet, and the slide valve cut off the steam at $\frac{1}{3}$d from the end of the stroke, and the cover on the exhausting side of the valve be $\frac{1}{8}$th of the valve stroke. In a line with $\frac{1}{8}$th, and under the columns B C, will be found ·178 and ·033, which, multiplied respectively by 72 inches, will give 12·8 inches for the distance of the piston from the end of its stroke, when the exhausting port before it is shut, and 2·38 inches when the exhausting port behind it is opened.

Temperature and elastic Force of Steam in Atmospheres.			
Force.	Temp.	Force.	Temp.
1	212°	13	380·66°
1½	234°	14	386·94°
2	250·5°	15	392·86°
2½	263·8°	16	398·48°
3	275·2°	17	403·83°
3½	285°	18	408·92°
4	293·7°	19	413·78°
4½	300·3°	20	418·46°
5	307·5°	21	422·96°
5½	314·24°	22	427·28°
6	320·36°	23	431·42°
6½	326·26°	24	435·56°
7	331·7°	25	439·34°
8	341·78°	30	457·16°
9	350·78°	35	472·73°
10	358·78°	40	486·59°
11	366·85°	45	499·14°
12	374·0°	50	510·6°

Expansion of Air by Heat in Degrees of Fahrenheit.			
Temp.	Vol.	Temp.	Vol.
32°	1000	100	1152
33	1002	110	1173
34	1004	120	1194
35	1007	130	1215
36	1009	140	1235
37	1012	150	1255
38	1015	160	1275
39	1018	170	1295
40	1021	180	1315
45	1032	190	1334
50	1043	200	1364
55	1055	210	1372
60	1066	212	1376
65	1077	302	1558
70	1089	392	1739
75	1099	482	1919
80	1110	572	2098
90	1132	680	2312

A	B	C	D	E	F	G	H	K
1-8th	·178	·033	·161	·026	·143	·019	·126	·012
1-16th	·130	·060	·118	·052	·100	·040	·085	·030
1-32d	·113	·073	·101	·066	·085	·051	·069	·042
0	·092	·092	·082	·082	·067	·067	·055	·055
A	**L**	**M**	**N**	**O**	**P**	**Q**	**R**	**S**
1-8th	·109	·008	·093	·004	·074	·001	·053	·001
1-16th	·071	·022	·058	·015	·043	·008	·027	·002
1-32d	·053	·033	·043	·023	·033	·013	·024	·004
0	·043	·043	·033	·033	·022	·022	·011	·011

Length of Stroke of the Valve in Ins.	Amount of Cover required on the Steam Side of the Valve to cut the Steam off at any of the under-noted parts of the Stroke.							
	$\frac{1}{3}$	$\frac{7}{24}$	$\frac{1}{4}$	$\frac{5}{24}$	$\frac{1}{6}$	$\frac{1}{8}$	$\frac{1}{12}$	$\frac{1}{24}$
24	6·94	6·48	6·00	5·47	4·90	4·25	3·47	2·45
23	6·65	6·21	5·75	5·24	4·69	4·07	3·32	2·34
22	6·36	5·94	5·50	5·02	4·49	3·89	3·13	2·24
21	6·07	5·67	5·25	4·79	4·28	3·72	3·03	2·14
20	5·78	5·40	5·00	4·56	4·08	3·54	2·89	2·04
19	5·49	5·13	4·75	4·33	3·88	3·36	2·74	1·94
18	5·20	4·86	4·50	4·10	3·67	3·19	2·60	1·83
17	4·91	4·59	4·25	3·88	3·47	3·01	2·45	1·73
16	4·62	4·32	4·00	3·65	3·26	2·83	2·31	1·63
15	4·33	4·05	3·75	3·42	3·06	2·65	2·16	1·03
14	4·05	3·78	3·50	3·19	2·86	2·48	2·02	1·43
13	3·76	3·51	3·25	2·96	2·65	2·30	1·88	1·32
12	3·47	3·24	3·00	2·74	2·45	2·12	1·73	1·22
11½	3·32	3·10	2·87	2·62	2·35	2·03	1·66	1·17
11	3·18	2·97	2·75	2·51	2·24	1·95	1·58	1·12
10½	3·03	2·83	2·62	2·39	2·14	1·86	1·51	1·07
10	2·89	2·70	2·50	2·28	2·04	1·77	1·44	1·02
9½	2·65	2·56	2·37	2·17	1·93	1·68	1·32	·96
9	2·60	2·43	2·25	2·05	1·84	1·59	1·30	·92
8½	2·46	2·29	2·12	1·94	1·73	1·50	1·23	·86
8	2·31	2·16	2·00	1·82	1·63	1·42	1·15	·81
7½	2·16	2·02	1·87	1·71	1·53	1·33	1·08	·76
7	2·02	1·89	1·75	1·60	1·43	1·24	1·01	·71
6½	1·88	1·75	1·62	1·48	1·32	1·15	·94	·66
6	1·73	1·62	1·50	1·37	1·22	1·06	·86	·61
5½	1·58	1·48	1·37	1·25	1·12	·97	·79	·56
5	1·44	1·35	1·25	1·14	1·02	·88	·72	·51
4½	1·30	1·21	1·12	1·03	·92	·80	·65	·46
4	1·16	1·08	1·00	·91	·82	·71	·58	·41
3½	1·01	·94	·87	·80	·71	·62	·50	·35
3	·86	·81	·75	·68	·61	·53	·44	30

Screw Steam Vessels.

TABLE showing the principal Dimensions and the Performance per Experiments of the Screw Steam Vessels, "Free Trade," "Eider," "Thames," and "Elbe," constructed by Boulton, Watt, & Co., 1847.

	"Free Trade."	"Eider."	"Thames."	"Elbe."
Length between perpendiculars.	140·0 ft.	140·0 ft.	130·0 ft.	130·0 ft.
Breadth	25 ft. 8 in.	25 ft. 8 in.	24 ft. 2 in.	24 ft. 2 in.
Depth	14 ft. 6 in.	14 ft. 6 in.	13 ft. 6 in.	13 ft. 6 in.
Tonnage	$424\frac{54}{94}$.	$424\frac{54}{94}$.	355	355
Immersed sectional area	166 sq. ft.	179 sq. ft.	136 sq. ft.	133 sq. ft.
Draft of water	8 ft. 4 in.	8 ft. 10 in.	7 ft. 4½ in.	7 ft. 2 in.
Speed in knots	7·66	7·31	7	– –
" miles	8·9	8·32	8·209	7·544
Diameter of screw	7·9 ft.	8·3 ft.	7·3 ft.	7·3 ft.
Slip of the screw in miles	1·33	– –	– –	– –
Pitch of screw	9·9 ft.	10·3 ft.	8·0 ft.	9·3 ft.
Length of screw	1·3 ft.	1·8 ft.	1·4 ft.	1·4 ft.
Number of blades of screw	Three.	Two.	Two.	Two.
Number of revolutions of screw per minute	94½	90	108	93
Diameter of cylinder	31½ in.	31½ in.	27 in.	27 in.
Length of stroke	3 ft.	3 ft.	2·6 ft.	2·6 ft.
Number of cylinders	Two.	Two.	Two.	Two.
Number of strokes per minute	31½	30	36	31
Vacuum by the barometer	27 in.	26 in.	24½ in.	26 in.
Nominal power	60 horses.	60 horses.	40 horses.	40 horses.
Actual power	132 horses.	146 horses.	102 horses.	100 horses.
Pressure of steam in boiler	13 lbs.	13 lbs.	12 lbs.	11 lbs.
Average effective pressure on piston	15·75 lbs.	17·30 lbs.	16·975 lbs.	19 lbs.
Steam cut off at — of stroke	⅓	⅓	¼	¼
Cubic feet of water evaporated per hour	105 ft.	105 ft.	70 ft.	70 ft.
Square feet of surface of boiler to evaporate a cubic foot	9·5 ft.	9·5 ft.	9·585 ft.	9·585 ft.
Square feet of fire bars to evaporate a cubic root	0·433 ft.	0·433 ft.	0·4928 ft.	0·4928 ft.
Actual horse-power produced by the evaporation of a cubic foot.	1·257 h. p.	1·39 h. p.	1·457 h. p.	1·428 h. p.
Sectional area of tubes per cubic foot evaporated	8·51 sq. in.	8·51 sq. in.	10 sq. in.	10 sq. in.
Multiple of the wheels	1 to 3	1 to 3	1 to 3	1 to 3
Diameter of wheel	9 ft. 4½ in.	9 ft. 4½ in.	8 ft. 6 in.	8 ft. 6 in.
" pinion	3 ft. 1½ in.	3 ft. 1½ in.	2 ft. 10 in.	2 ft. 10 in.
Pitch	3·927 in.	3·927 in.	3·141 in.	3·141 in.
Breadth	2 teeth, each 6 in.	2 teeth, each 6 in.	2 teeth, each 4½ in.	2 teeth, each 4½ in.
Whether teeth of wood	Wheel geared with horn-beam. The pinion of iron.	Wheel geared with horn-beam. The pinion of iron.	Wheel geared with horn-beam. The pinion of iron.	Wheel geared with horn-beam. The pinion of iron.

INDEX.

www.ingramcontent.com/pod-product-compliance
Lightning Source LLC
LaVergne TN
LVHW010225110826
845151LV00004B/1184

* 9 7 8 1 4 2 5 5 2 7 2 2 8 *